J. G. (John George) Wood

**Horse and Man**

Their Mutual Dependence and Duties

J. G. (John George) Wood

**Horse and Man**
*Their Mutual Dependence and Duties*

ISBN/EAN: 9783744648264

Printed in Europe, USA, Canada, Australia, Japan

Cover: Foto ©berggeist007 / pixelio.de

More available books at **www.hansebooks.com**

*THEIR MUTUAL DEPENDENCE*

*AND DUTIES*

BY THE

REV. J. G. WOOD

AUTHOR OF

'HOMES WITHOUT HANDS' 'BIBLE ANIMALS' ETC

LONDON
LONGMANS, GREEN, AND CO.
1885

# PREFACE.

ALTHOUGH this work treats of the Horse, it is not 'horsey,' because I am not a horsey man. Horsiness has no place in it, and, for a time at least, will probably be in opposition to it. I only ask the reader to bear in mind that I pretend to no discoveries, and advance no theories. I simply state facts, offer evidence of those facts, adduce proofs of that evidence, state how and where these proofs can be verified, and then leave the impartial reader to draw his own conclusions.

# CONTENTS.

### CHAPTER I.

The hoof as distinct from the foot—Dependence of one part on the others—Skeleton of the horse—Story of 'Orlando'—Use of scientific language—The spine and the limbs—The legs of the horse compared with the limbs of man—Gradual development of the horse—The anchitherium and hipparion—Structure of the fore-limb, or arm—Rotation of the fore-arm prevented—Small size of the original horse—An eight-hoofed horse—Hind legs of the horse compared with the legs of man—The true knee of the horse—Elasticity of the structure . . . 1

### CHAPTER II.

Connection between the foot and the hoof—Extent of the foot—What is horn?—Original dwelling-place of the horse—Climbing powers of the horse—Requisites of the hoof—STRENGTH, because it has to bear the weight of so heavy an animal—LIGHTNESS, lest the horse should have to lift needless weight—The 'coffin' bone and its structure—HARDNESS, to endure contact with bad ground—SHARPNESS, round edge to enable the horse to climb—CLINGING, to suit smooth and slippery ground—SELF-REPAIRING, for restoration of worn material: the Wall, Frog, and Sole—Structure of the WALL—The horny laminæ—Their origin and mode of growth—Over-grown hoofs—Elasticity begins with angle of pastern—The FROG, its position, structure, and office—Analogies between Nature and human inventions—The SOLE: its structure and double office . 22

## CHAPTER III.

Internal structure of the hoof—The hoof compared with the Jacquard loom—Longitudinal section of pastern—The 'navicular' bone: its origin, form, and office—The tendons—Flexor and extensor muscles and their tendons—The 'coronary' ring and its object—The hoof of the horse and the nail of man—The 'quick' of the nail—The 'sensitive' or 'vascular' laminæ—Mutual dependence of the horny and vascular laminæ—Analogy of laminæ of whalebone and those of hoof—Expansive property of hoof—Mr. Miles's experiments—Advantage of this property in a hunter—Effect of shoes in leaping—In any pace the heel comes on the ground before the toe—Importance of this fact . . . . . . . . . . . . 38

## CHAPTER IV.

Expansion acknowledged, acted upon, and perverted—Authorised directions for shoeing—'Thinning' and 'opening'—Horse-shoeing in England and America—'Dew-drops,' *i.e.* oozing blood—The diseases called by the name of 'thrush'—True causes of thrush—Supply of blood to hoof—Nature's balance of supply and waste—Office of the blood—'Thrush' a safety-valve for inflamed blood—Derivation of the word 'founder'—'Thrush' never seen in wild horses—A comprehensive challenge—Where do wild horses keep their knives?—The frog again—Lieutenant Douglas on the frog and its value. Mr. T. Gepp's experience . . . . . . . . . 56

## CHAPTER V.

The FROG continued—'Hammering' on the roads—Cause of the hammering—King Theodore's horse 'Hammel'—Elasticity of the frog—The 'Village Blacksmith'—The bicycle wheel—Analogy between carriage springs and the horse's hoof—Lodgment of stones—Mr. S. Chapin and his bare-footed horse—Hartmann's safety pad—Spiked shoes—American shoeing—Injury to the untouched frog impossible—'Greasy' roads—Manchester 'lurry' horses—The streets of Manchester—A doctor's horses—Shoes of lurry horses—Ludgate Hill and its terrors—Lieut. Douglas's testimony—Indiarubber soles on ice and on board ship—The hoof an organ of touch—Mr. J. Bellows's story—'Free Lance's' view of the frog and its uses—Relative dependence of the frog and navicular bone . . 69

## CHAPTER VI.

Horseshoeing on 'improved principles'—Hot-fitting—'Clips' and their origin—Groove cutting in the hoof—Natural thatch of the coronet—Natural varnish of the hoof—Abuse of the rasp—Blacking hoofs—Effect of grease on the hoof—'Stopping' hoofs—The 'bottle of oils'—Its effect on the hoof—Drugs in stables—Horses poisoned—Thirty thus lost by one owner—Anti-drug Association—The rashness of ignorance . . . 86

## CHAPTER VII.

The shoe—Artificial roads and artificial protection—Variety in roads—Straw shoes of Japan—Raw hide or 'parflèche' shoes of North American Indians—Shoe nails—Their ordinary size and number employed—Diminishing thickness of wall—An old Scotch law—The 'unilateral' system—A hoof prepared on the 'improved system'—A mangled hoof restored by Nature—The dangers of shoe nails—Cut nails and forged nails—A remarkable accident—Effect of a heavy shoe on the horse—'Marden' and the dead heat—Effect of a heavy shoe on the muscles—Lancashire clogs and French *sabots*—Cetewayo and followers in England—The 'lurry' horses and their shoes—Lieut. Douglas's calculations—Loosened hoofs . . . 101

## CHAPTER VIII.

The calk, or calkin—Horses on pattens—Two strange accidents—Calks in America—Supposed uses of the calk—Mr. Bowditch's testimony—Weight thrown on the edge of the coffin bone—High-heeled boots and their effects—The battle of the shoes—Recognition of defects in shoeing—The Goodenough shoe and its object—Jointed shoes—The Clark jointed shoe—The screw shoe—Expansion and contraction—The effect of the screw on the hoof—Tips, and how to fasten them—The Charlier, or 'pre-plantar' shoe—How to apply it—Usually too large and in danger of breaking and twisting—Best length and weight for a Charlier shoe—Man *versus* Nature—A series of happy thoughts—Their results upon each portion of the hoof . . 116

## CHAPTER IX.

The shoe useful in proportion to its lightness—Therefore, the best shoe seems to be none at all—Capability of the human foot—Value of an army dependent on its marching power—Lord

Wolseley's axiom—Edinburgh lasses—Moccasin *versus* boot—
Mansfield Parkyns in Abyssinia—Ladies and children at the
sea-side—Charles Waterton in Guiana—Col. Dodge's account
of the North American Indian's pony—A race between the
Indian's pony and the high-bred horse—Exmoor and Dartmoor
ponies—Description of these 'moors'—How to make a horse's
hoof tender—The hoof an organ of all-work—Saddle and
draught—Col. Burnaby's opinion . . . . . . . 142

## CHAPTER X.

Unshod horses now at work—Dr. Llewellyn's horse on London
roads—Thirteen thousand miles without shoes—Always went
lame when shod—His transfer to Mr. A. F. Astley—Mr.
Astley's horse 'Tommy'—State of his hoofs when bought—
Process of training—Work done by him unshod on hard
roads—Photographs of his hoofs—Mr. Whitmore Baker's
mare 'Stella'—Facts *versus* theory—Photographs of 'Stella'
and her hoofs—Letters from Mr. Baker—Work done by
'Stella' barefooted—Galloping over ice or loose stones—Mr.
Baker's offer to enable other horses to work unshod—His
preparation for hoofs and its possible value—Influence of
external conditions on the hoof—The condemned tramcar
horse—Result of removing the shoes—Five hundred and forty
miles unshod—Photograph of the hoof in transitional state—
Mistaken benevolence . . . . . . . . 158

## CHAPTER XI.

Hardening and renovating the hoof—Variety in hoofs—Thrush
concealed by shoe—Dr. Brierley's horses—Horses in Italy—
Mr. Theodore E. Williams's horse 'Prince'—Lame when shod
—Experiment on another horse, and result—Mr. Herbert
Smith's experiments—Altered shape of hoof—Need of perseverance—Xenophon's rules for hardening the hoof—General
summary of the subject . . . . . . . . 178

## CHAPTER XII.

The Professional Eye—Fashion and nature—The curb—Weight and
size of bit—The BEARING-REIN—Three kinds of bearing-rein—
The gag bearing-rein—Mechanical parallel—The over-head
rein—Neck of the horse—Great ligament of the neck and its
attachments—Vertebræ of neck and spine—Vertebræ and

railway buffers—Arrangement of a train—The martingale—Rattling of harness and tossing of heads—Sir Arthur Helps' opinion—Effect of the gag bearing-rein on the spine and feet—The 'burr' bit of America—Mr. Henry Bergh's work—The locomotive and the horse . . . . . . . 198

## CHAPTER XIII.

The bearing-rein continued—The locomotive and the brake—Probation of an engine driver—The bearing-rein and the break—Leading reins converted into bearing-reins—Railway companies and the bearing-rein—Theories as to the bearing-rein—Its supposed use in preventing the horse from falling—Bearing-reins and hills—Harness in Scotland—The bearing-rein in Bristol—Mr. Cracknell's testimony—The bearing-rein and runaway horses—A grievous experience—The shoulder injured by the bearing-rein—Testimony of more than a hundred veterinary surgeons—'Roaring' caused by the bearing-rein—Mistaken zeal—Summary—'Free Lance's' check for a runaway horse . . . . . . . . . . . 222

## CHAPTER XIV.

The BLINKER and its supposed uses—Cropping of ears—Alleged necessity for cropping—Contradictory temperament of the horse—Courage and timidity—Inquisitiveness—Rarey's principle—The kettle-drummer's horse—Mr. C. H. Tamplin's experience—Obstinacy of a coachman—Value of the 'master's eye'—Waterton at Walton Hall—Letter from 'C. F. W.' to the *Field* newspaper—'Jockeying' adviser—Modified blinkers—Eye of the horse—Cruel superstitions—The third eyelid, or 'nictitating membrane,' and its use—The groom's rashness and its effects . . . . . . . . . . . 244

## CHAPTER XV.

The mane and the practice of 'hogging'—The tail and its office—What man does to the tail—Docking—A puzzled J.P.—The professional eye again—Docking and lock-jaw—Nicking—An unexpected ally—'Conducive to human safety'—The tail and the crupper—Winter and summer coats of the horse—Clipping and singeing—American horses in winter—Fashion with man and horse—The groom's real reason for clipping—Mayhew and Lupton's opinions of clipping—The 'moulting' of birds . . 264

## CHAPTER XVI.

The lungs of the horse and their comparative size—Their shape and position—Their demand for air—Defective stables—Ventilation—A shining coat—The stomach of the horse and its small size—Comparison with the stomach of the ox—Mistakes in feeding—Result of over-loading the stomach—Experience of a veterinary surgeon—Water, when to give, and how much—Traditions of trainers—A Turkish custom—Purity of water and water vessels—Sloping floors and their evils—The locomotive and the horse again—A sloping couch—The manger and drinking trough—Structure of the head and throat—The stable door—Width of stalls—'Weaving' and crib-biting—The electric manger . . . . . . . . . 287

## CHAPTER XVII.

The horse and the locomotive again—'Vice,' in horses and its invariable cause—Mayhew's opinion—Vice in cavalry horses—The soldier and the 'irreclaimable' savage—New mode of treatment—Apparent failure and ultimate success—A relapse when in strange hands—Another 'irreclaimable' savage—Story of 'Fly,' 'The Baroness' and 'War Eagle,' all three being New Zealand horses—The horse's capacity for affection—Its peculiar love for man—The horse a gregarious animal—Bulgarian horses—'Spoiled' horses—The horse's desire to obey man—A circus horse—Gilpin redivivus—Cavalry horses and their habits of obedience—The old horse at a review—Disbanded horses in a thunderstorm—The 14th Hussars at the Cape—Escape of their horses—An amateur review without officers—Muster of loose and wounded horses after battle—Mr. Luck's horse—Imprisoned in winter—Endurance of the horse—'Sam' and his tricks . . . . . . . . 306

# INTRODUCTION.

As the reader may have inferred from the Preface, this work has nothing to do with the HORSE as an instrument of sport. Neither will anything be found in it with relation to the breeding of horses, or with their technical 'points;' with buying and selling, with the tricks of trade (which, by the way, are not one whit worse than the tricks of any other trade), nor with medical and surgical treatment.

Let the breeding of horses be left to those whose long experience (tempered with a little fresh blood from an outsider) enables them to supply from the same original source the elephantine cart horse, the light and swift racer, the sturdy cob for general service, the pony for our young people, or that wonderful combination of speed and power which carries its rider to the front in the hunting field.

Let the horse trade be left to those who understand the ever-shifting vagaries of fashion, and who are familiar with the various 'points' which give a fictitious value to the horse as they do to the dog, the canary, the rose, the dahlia, or the tulip.

Let the medical and surgical treatment of the horse be left to those who have been trained in an acknowledged veterinary college, and who can produce the diploma which testifies to their scientific and practical capability. Nowhere do I presume to instruct the veterinary surgeon. On the contrary, I urge, throughout the work, that no one ought to be allowed to administer the mildest of drugs, or to perform the slightest of operations, unless he be legally qualified to do so.

Throughout the work I draw a parallel between the horse and the steam-engine, and try to show that those who have the management of the former or the latter will be adapted to their task in proportion to their knowledge. It is not needful that this knowledge should comprise those details which belong to pure science.

There is not the least necessity that an engine-driver should be learned in the chemical constituents of the metals of which the engine is constructed, of the fuel which is consumed, or of the water which is converted into steam.

But it is necessary for an efficient engine-driver to know the amount of strain which iron, steel, bell-metal, and brass can bear, and why these metals should be used in their respective places. It is necessary that he shall be able to distinguish the different qualities of coal, and to know how much steam can be raised by a given quantity of fuel.

'Engine and man,' in fact, must go together, and so must 'horse and man.' Therefore, the reader will not find his time wasted by anatomical minutiæ which do not bear on the practical management of the horse, and I have contented myself with slightly describing the general structure of the animal, and giving such details of the foot, hoof, neck, lungs, and stomach as ought to be known by every one who possesses, or has the management of, a horse.

In civilised countries, horse and man are inseparable. Neither can exist without the other, and each owes a duty to the other. Take any large city, deprive the horse of the services of man, and in a fortnight at the most not a horse would be left alive. In the same city, deprive the man of the services of the horse, and he would lose much of his civilisation, being forced to work with his body instead of his mind.

There is, however, one fundamental distinction

between the two beings. Man has his choice of action, whereas the horse has none, so that the chief responsibility is thrown upon man. Man can teach the horse, but the horse, unfortunately for itself, cannot teach man.

Each owes a duty to the other.

Man is bound to afford to the horse the food and shelter which it needs, and which it cannot procure without his aid. On the other side, in return for its means of existence, the horse is bound to give to man the benefit of its labour. In the following pages I have endeavoured to show how horse and man can be fellow-workers instead of master and slave, how the life of the horse can be prolonged as much as possible, and how the animal can be enabled to do the maximum of work during its lifetime.

# HORSE AND MAN.

## CHAPTER I.

The hoof as distinct from the foot—Dependence of one part on the others—Skeleton of the horse—Story of 'Orlando'—Use of scientific language—The spine and the limbs—The legs of the horse compared with the limbs of man—Gradual development of the horse—The anchitherium and hipparion—Structure of the fore-limb, or arm—Rotation of the fore-arm prevented—Small size of the original horse—An eight-hoofed horse—Hind legs of the horse compared with the legs of man—The true knee of the horse—Elasticity of the structure.

WE will begin with the foot. This may seem simple enough, but it is really a question of great complexity.

In the first place I suppose that at least ninety-nine persons out of a hundred take for granted that the hoof and the foot are identical, and that the former is a solid lump of hard horn, upon which an iron shoe can be nailed, in order to protect it against artificial roads. They also have a vague idea that the hoof is an isolated portion of the horse's frame, and has no definite connection with any other part.

Now, in the first place, the hoof is not the foot, but only the extremity of a portion of the foot. In the second place, instead of being a lump of horn, it is one of the most elaborate pieces of mechanism in the whole of the animal kingdom. In the third place, it is an integral portion of an entire and symmetrical organisation; each part being connected with and dependent upon all the others, and all subservient to the great object of procuring food.

No one part could be altered without changing all the others. To put an extreme case:—Suppose that the teeth of the horse were exchanged for those of the lion, the hoof would be useless, as it could not secure living prey. The stomach would be useless, because it would be incapable of digesting raw flesh, and the teeth, from their structure, would be unable to masticate, and therefore could not chew grass.

Conversely, if the hoof were exchanged for lion's paws and talons, they would be quite unsuited to the pastures in which the horse finds its food, and so the teeth would have to be altered.

Now let us see what is the general structure of the horse, and what relationship is borne to it by the hoof. The accompanying illustration represents the bony framework of the horse. It was drawn from the skeleton of the celebrated racehorse

'Orlando,' who won the Derby in 1844, the year of the famous 'Running Rein' swindle.

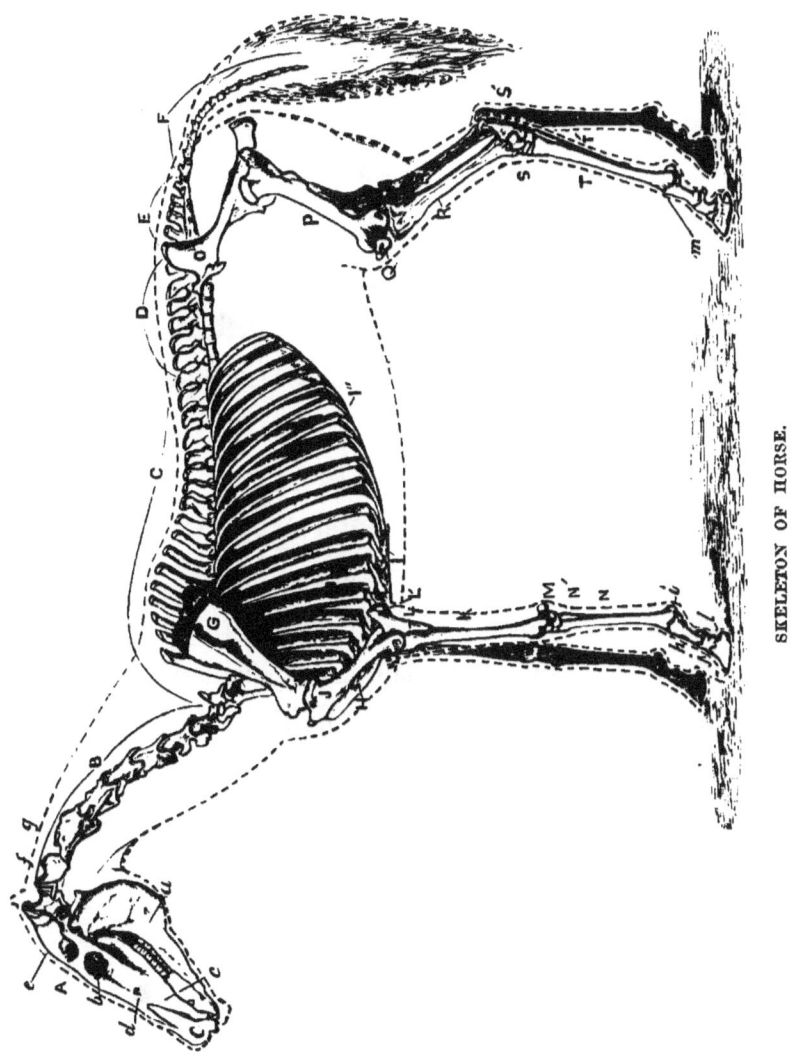

As the details of this race may not be familiar to the present generation, I will briefly mention them.

Most people know that the 'Derby' is intended for three-year-old animals, although a French writer does tell us that a certain horse won the Derby three times successively. Among the horses which were entered was one called 'Running Rein.' For that animal, a four-year old horse called 'Maccabæus' was substituted under the same name, and won the race. The fraud, however, was discovered, and great scandal arose in consequence. The result was that 'Maccabæus' was disqualified, and the race was awarded to 'Orlando,' who ran second. (Parenthetically I take the opportunity of wishing that two years could be added to the age, a mile to the course, and considerable addition made to the weight.)

The skeleton is in the Museum of the Royal College of Surgeons, and, like many other figures in this work, was drawn by the kind permission of Professor W. H. Flower, of the British Museum, who was then curator of the wonderful museum in Lincoln's Inn Fields, and to whom I return most cordial thanks for his kindly assistance.

We will first take a general survey of the skeleton, and then review each part in detail.

I shall in the course of this work avoid as far as possible the use of purely scientific language. It is not written for pupils of the Royal Veterinary College, nor for persons skilled in comparative anatomy. It is intended for the use of all those who possess or who have charge of horses, and therefore will be couched in language which can be understood by all persons of ordinary education.

Scientific terms, however, cannot be entirely ignored in anatomy any more than in other subjects.

For example, it would be impossible to give the most superficial idea of the carpenter's art without employing such terms as 'mortice,' 'tenon,' 'rabbeting,' 'mitre,' and a hundred other technical words. Similarly, nautical terms are quite unintelligible to an ordinary landsman, and yet it would be impossible to work a ship, or even to sail a five-ton cutter, without knowing the meaning of such expressions as 'luff,' 'jibe,' 'topping lift,' 'port,' 'starboard,' 'wear,' 'in stays,' &c.

So, in describing the structure of the horse, many technical words must be employed, but in every case they will be fully explained when first used.

---

I WILL now ask the reader to examine the skeleton in general. All vertebrates are constructed on the same principle; modified, or 'differentiated' in detail according to the life which the animal is intended to lead. The only essential skeleton of a vertebrate animal consists of a skull followed by a series of vertebræ.

Limbs are only appendages; many vertebrates, such as the generality of the snake tribe, not possessing them. Even ribs are not necessary for existence, as we see is the case with the frogs, in which animals the ribs are totally absent. So, following the order of nature, we will first take the skull and vertebræ.

A represents the skull itself, *b* is the eye-hole (*orbit*), *c* is the upper jaw (*superior maxillary*), *d* are the nasal bones—*i.e.* the bones of the nose, *e* is the lower jaw (*inferior maxillary*), *e* is the ear-hole (*auditory foramen*).

As to the teeth, the reader will observe that there is a considerable gap between the canine teeth, or ' tusks,' and the molars, or grinders. Were it not for this gap, the bit could not be inserted in the mouth, and man would lose his most powerful means of guiding the animal.

B comprises the neck bones (*nuchal* or *cervical vertebræ*). These are always seven in number in the mammalia, no matter what the length of the neck may be. Whether the neck may belong to the giraffe, which feeds on tree-leaves eighteen feet from the ground, or whether it be the property of the whale, which has no perceptible neck at all, the nuchal vertebræ are seven in number. In the former animal they are greatly elongated, while in the latter they are flattened and fused together. Those of the horse are intended to give flexibility to the neck, and to assist in 'clothing it with thunder.'

The vertebra marked $f$ is called the 'atlas,' because in man this vertebra supports the head, as the mythical giant Atlas was said to support the earth; while $g$ is called the 'axis,' because it is so constructed as to enable the animal to turn its head.

These seven vertebræ are represented on a larger scale in the illustration on page 8. The reader will observe that the bones are furnished with projections. These are intended partly for the attachment of the muscles, and partly for the branches of the great ligament that runs over the back of the neck and supports the head. This ligament and its attachments will be seen when we treat of the neck in connection with the bearing rein.

Next come the eighteen vertebræ of the back, or 'dorsal' vertebræ, each having a long flattened process. The reader will notice that some of these processes lean backwards, while others are directed forwards. Between them are several which are upright, and upon them is the proper place for the saddle.

Then there are the six vertebræ of the loins (*lumbar vertebræ*), followed by six vertebræ called 'sacral.' They are fused together, and are termed the 'sacrum.' To this portion of the spine the hip, or 'pelvic,' bones are attached.

VERTEBRÆ OF NECK.

Lastly, come the vertebræ of the tail, or 'caudal' vertebræ. These are rather variable in number, but on the average are fifteen.

Now we will take the appendages of the vertebræ.

In front of the body is the breast-bone, or 'sternum.' The ribs are divided into two sets. First come the 'true ribs,' which are connected directly with the sternum, and then come the 'false ribs,' which are indirectly connected with the sternum by gristle (or cartilage).

Lastly we come to the limbs. They are formed of the same bones as those which constitute the arms and legs of man, but as they are intended to perform different offices, the bones are greatly modified in structure.

We will begin with the fore-leg, which is analogous to the arm of man. First comes the shoulder-blade (*scapula*), which, except that it is much longer in proportion to its width, is not very different from that of man. But, in proportion as we pass from the shoulder to the toe, we shall find the modifications of form more and more strongly marked. So, the upper arm-bone (*humerus*) is very different from that of man.

With man, the whole weight of the body is supported on the hinder limbs, the arms being set free for particular purposes. The humerus of man, therefore, is long, and comparatively slight. But such a bone would be absolutely useless in a horse, where it has to bear the weight of the fore-part of the body, plus the heavy neck and head, and would inevitably be shattered at the first leap.

Therefore, the humerus of the horse (see J in the figure) is short, thick, and set in a sloping direction, so as to avoid a direct shock as the animal alights from its leap. The reader will notice that the lower end of this bone is enlarged, expanded, and very deeply grooved.

Next come the two bones of the fore-arm, K being the large bone (*radius*), and L the small bone (*ulna*). Both terms are Latin, the former signifying the spoke of a wheel, and the latter an arm.

Here is a bolder modification than in the humerus.

It is necessary for man that he should be able to rotate the hand, and therefore the two arm-bones are free at their extremities, and so constructed that they can partially roll over each other. But, for the hand of the horse to be capable of rotation would be an element of weakness, and therefore the bones are so modified that they can only be moved directly backwards and forwards. The ulna, therefore, is very much reduced in size, and does not

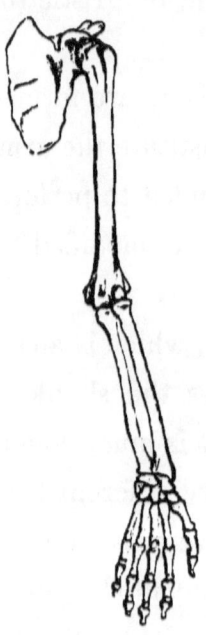

LEFT ARM (HUMAN.)

nearly reach to the lower end of the radius, to which it is fused so as practically to form a single bone. The upper end is prolonged, flattened, and plays in the groove at the lower end of the humerus, so that no rotation is possible. The prolonged portion forms the elbow of the horse (*olecranon*).

In order to show this structure more clearly, the joint is given on a much larger scale on page 12. Here A represents the lower end of the humerus with its double head and central groove; B is the radius, and C the ulna. The reader will here see how it is fused into the radius, and how its flattened prolongation plays up and down in the groove. D is the olecranon, or elbow.

Descending still lower, we find the bones of the wrist (*carpus*), or 'knee' as it is popularly called. These are much the same as those of the human wrist, except that they are simpler in structure.

Now, however, we find a most radical change. Compare the bones of the human hand with those of the horse. No resemblance seems at first sight to exist between them; yet we know that, different as they may seem, they are composed of the same elements. In order to solve this problem we must have recourse to geology and comparative anatomy.

The former science shows us that the horse of the present period is the last of a long series, extending through successive geological epochs and undergoing determinate modifications, so as to bring it in accord with its surroundings. Five or six links of this chain have already been discovered, and in all probability others will come to light. I shall only mention three of these links.

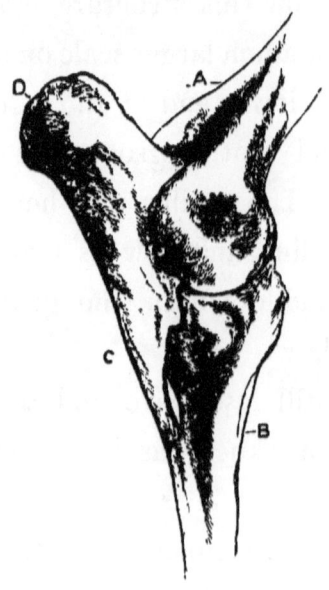

ELBOW, NEAR FORE-LEG (SEEN FROM RIGHT SIDE).

The reader may have been struck with the fact that the number of toes on the foot of a mammal is exceedingly variable, but yet has certain limits. Some, like man, have five toes or fingers (*phalanges*) on each extremity, while others, like the horse, have only one. Except by occasional monstrosity there are never more than five phalanges, while the animals sometimes have two, three, or four, as the case may be. In some, as in the tapirs, there are four toes on the front feet and only three on the hind limbs.

A comparison of the hand of man with that of the horse seems quite absurd, and yet it becomes simple and intelligible when systematically carried out. The

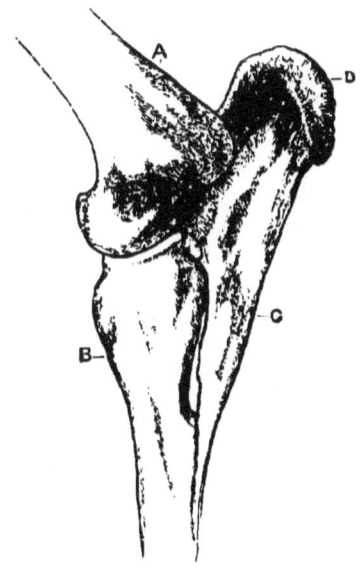

ELBOW, NEAR FORE-LEG (SEEN FROM LEFT SIDE).

reader must carefully keep in mind the fact that the so-called knee of the horse is really the wrist, and that the rest of the limb is really the modified hand.

The earliest horse known to geologists had five toes or fingers. It was quite a little creature, scarcely larger than an ordinary terrier dog. Then there came an animal in which the characteristics of the horse are much more strongly developed. It is appropriately named the Anchitherium—*i.e.* the creature approaching the horse. In this animal there were apparently three toes, all resting on the ground, and the other two almost touching it. Next came the Hipparion—*i.e.* an animal almost a horse.

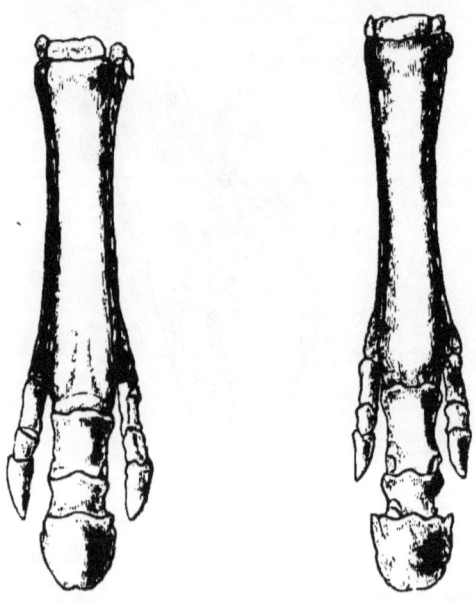

LEFT FOOT OF ANCHITHERIUM.   LEFT FOOT OF HIPPARION.

This creature had still the three toes, but only the central toe rested on the ground, the others being drawn upwards.

I intentionally used the word 'apparently' when describing the toes of the anchitherium, and for the following reason. Let the reader again look at the bones of a man's hand. Starting from the wrist are five bones, called the 'metacarpals'—*i.e.* following the carpus, or wrist. Suppose we call the thumb (as it really is) the first finger, we find that in many animals the third and fourth metacarpals are fused together, either partially or completely. In the modern horse they are fused (*anchylosed*) so completely, that it is not easy to detect the fact that they are really two bones. But, in the hipparion the line of junction is manifest, and still more so in the anchitherium. A similar line of junction may be seen in the very familiar bones of the oxen and sheep.

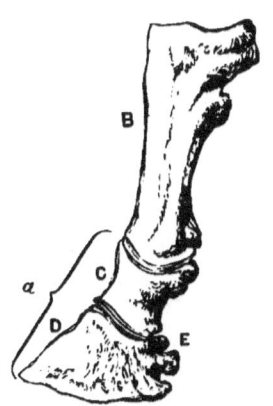

FINGER OR PASTERN BONES.

In these latter creatures, however, the third and fourth phalanges diverge, so as to produce the cloven hoof, but in the Horse tribe the phalanges, as well as the metacarpals, are fused together, so as to form a series of single bones, as here shown. Here B represents the first of the three finger joints, and is popularly called the Long Pastern bone. C is the middle joint, and is called the Short Pastern, while the last, or 'distal' joint, is termed the Coffin Bone. How these bones are connected with the hoof we shall see on a future page.

Now we can find a clue to the modification of the foot, or hand, whichever we like to call it. The long straight bone which reaches from the wrist to the pastern ('cannon,' or 'shank bone,' as it is sometimes called) is formed of the third and fourth metacarpals fused together. But what has become of the rest of the hand? Where is the thumb? What has been done with the second (index) and little (fifth) fingers? We can answer these questions by geology combined with comparative anatomy.

In every instance where the phalanges become rudimentary, the thumb is the first to be withdrawn. In the dog and cat, the thumb is familiar to all as the 'dew-claw.' Now, in the horse tribe the thumb was withdrawn even before the time of the anchitherium, leaving only three toes (or fingers) capable of resting on the ground—*i.e.* in the middle were the third and fourth toes fused into one, and on either side the index and little finger.

In the hipparion these external toes are partially rudimentary, so that though they still possess hoofs, they do not rest on the ground, and are as useless for walking purposes as those of the deer. In the true horse, as we now know the animal, these external toes have become perfectly rudimentary. There is no external trace of them, but when the skin is removed from the horse's shank two skewer-like bones are seen, one on each side of the cannon bone. These are the metacarpals of the index and little fingers, the phalanges having disappeared altogether, and the metacarpals themselves being reduced to mere splinters of bone, without a trace of a joint at either end.

I may here mention that even when the horse did appear, it was very unlike the beautiful animal of the present day. It was even smaller than the Shetland pony, had a head and neck very large in proportion to the body, a coarse and heavy mane, and was altogether a clumsy sort of animal. This we learn from the wonderful engravings upon bone or antler executed by those born artists the Men of the Caves. To them we are indebted for portraits of the mammoth, deer, horses, and other animals, drawn with a freedom, a truth, a vigour, and a fire of which our best animal artists might be proud.

The various breeds of horses which we possess at the present day are, so to speak, artificial, and are due to the constant influence of man. When deprived of this influence, they soon display a tendency to retrograde to the ancient type, their bodies and limbs gradually dwindling, but their heads remaining of their original size.

Every now and then a horse is born which exhibits traits of its ancestry, just as fancy rabbits of the purest lineage are apt to produce young which can hardly be distinguished from the semi-wild rabbits of our warrens. In point of size, the beautiful little pony 'Lady Jumbo,' which was exhibited in 1882, afforded a good example of the primitive horse. When shown at Islington she was only thirty inches in height, and was brought from the London Bridge Station to Islington inside a four-wheel cab.

Sometimes a horse is born with three toes on one foot. In the autumn of 1883, while staying in Boston, Mass., I saw a horse with eight hoofs, the second (i.e. the fore-finger) phalanges being almost as perfectly developed as the third and fourth. The supplementary hoof, although it did not quite reach the ground, was nearly as large as the actual hoof. Unfortunately —so are we swayed by custom—the owner had shod all the hoofs alike, a piece of barbarism which I lost no time in denouncing.

Now we will turn to the hind limbs. Here we find almost a repetition of the fore limbs, but the joints are differently arranged.

Instead of the blade bone there is the hip bone, or 'pelvis,' marked o in the illustration.

Then comes the thigh bone (*femur*) in place of the humerus. Here, again, a long and slender thigh bone would be a source of weakness to the animal, and so, like the humerus, it is shortened, thickened, and is set at a slope, so as to avoid a direct shock to it.

Next we come to the real knee of the horse—a joint which is professionally called the 'stifle,' why, I cannot imagine. This joint being the knee, there is a knee cap (*patella*), which performs the same office for the horse that it does for us. Dislocation of the patella is even more common with horses than with man, but fortunately can be reduced more easily, and does not leave such persistent weakness.

After the thigh bone come the two bones of the lower leg—the 'tibia' and 'fibula.' Both words are Latin; the former signifying a flute, and the other a buckle, or rather the tongue of a buckle. With us, they are popularly known as the large and small leg bones.

Next comes the ankle (*tarsus*), popularly called the 'hock.' As in the wrist, so in the ankle, the bones are not unlike those of men. The heel-bone, however, is longer, as it has to afford leverage to the great heel sinew or tendon (*tendo Achillis*), the severance of which is called 'houghing,' and lames for life either man or beast.

Just as in the fore-limb the metacarpals are rudimentary, with the exception of the third and fourth which are fused into a single bone, so it is with the 'metatarsals' of the horse—marked T in the illustration—which correspond with the bones forming the instep of man. Similarly, the phalanges, or toes, have totally vanished, except those which belong to the cannon bone, and, like the metatarsals, are fused into single bones.

LEFT LEG AND PELVIS (HUMAN).

Thus, from the so-called 'knee' to the end of the fore limb is the real hand of the horse, and from the hock to the end of the hind limb is the true foot, the horse walking on the tips of the third and fourth fingers of the hand, and on the tips of the third and fourth toes of the foot.

Before leaving the skeleton for the present, I wish to call the attention of the reader to the mode in which the joints of the limbs work.

The humerus is directed backwards, while the femur slopes forward. Consequently, when the animal moves, the elbow and stifle (the real knee), which are on a level with each other, bend towards each other. Then, the two next joints—*i.e.* the so-called knee and the hock—bend away from each other. The whole structure, together with the angle of the pastern, ensures that elasticity which is so necessary for the animal's welfare.

In order that these bones may be the better impressed upon the mind, I strongly recommend the reader to colour the figure, just as maps are coloured, for the purpose of making them more easily intelligible. My own figure is coloured as follows:—

Taking the same order as has been observed in the description, the skull is olive-green, and the lower jaw blackish-grey. The cervical vertebræ are yellow ochre, the dorsal vertebræ blue, the lumbar vertebræ red, the sacrum brown, and the tail red.

The true ribs are olive-green, and the false ribs purple. As for the limbs, the scapula and pelvis are pink, and the humerus and femur olive-green. The radius and tibia are yellow ochre, while the ulna and fibula are red. The carpus and tarsus are brown, the metacarpals and metatarsals purple, the splint bones scarlet, and the phalanges brown. Of course, the darkened bones of the right limbs should be left uncoloured.

## CHAPTER II.

Connection between the foot and the hoof—Extent of the foot—What is horn?—Original dwelling-place of the horse—Climbing powers of the horse—Requisites of the hoof—STRENGTH, because it has to bear the weight of so heavy an animal—LIGHTNESS, lest the horse should have to lift needless weight—The 'coffin' bone and its structure—HARDNESS, to endure contact with bad ground—SHARPNESS, round edge to enable the horse to climb—CLINGING, to suit smooth and slippery ground—SELF-REPAIRING, for restoration of worn material: the wall, frog, and sole—Structure of the WALL—The horny laminæ—Their origin and mode of growth—Over-grown hoofs—Elasticity begins with angle of pastern—The FROG, its position, structure, and office—Analogies between Nature and human inventions—The SOLE, its structure and double office.

HAVING now taken a comprehensive glance at the skeleton of the horse, and seen the relationship between the foot and the animal, we will go into a little more detail, and see where is the connection between the foot and the hoof.

The popular idea on this subject is that the hoof, or at all events the hoof together with the fetlock, is the foot; whereas, as we have seen, the foot of the fore-limbs begins at the so-called knee, and that of the hind limbs at the hock.

The hoof is, in fact, the nail of the finger or toe,

enlarged and modified to perform a series of offices. In material and position it is identical with the talon of the lion or tiger and the claw of the eagle. It may seem strange that the delicate pinky nail of a lady's finger should be one and the same with the hoof of the horse, but such is really the case.

Now for the structure of the hoof, or nail, of the horse.

Instead of being a mere lump of horn, it is a sort of horny case, or box, intended to protect the sensitive structure which it surrounds. The offices which it serves are many, though several of those offices are practically ignored by civilised man.

Indeed, even those who really know the structure of the horse have made the most curious mistakes about the work of the hoof. One writer states that the horse was intended to live on moist or even marshy land, whereas such ground is the very worst for the hoof, and is carefully avoided by the horse whenever it can find hard and dry ground to stand upon.

The animal always instinctively tries to find a hard surface on which to stand. In America the 'mustangers'—*i.e.* the men who get their living by catching and training the mustangs, or wild horses—invariably choose the hardest and stoniest places for the 'corrals,' in which they keep the horses until they are wanted.

Another writer says that the horse was intended to live on level pasture land.

Now, it is admitted that Central Asia is the original home of the horse, and that the animal is not very likely to find in those regions either marsh or pasture land. In point of fact, the horse is intended to suit a very wide range of locality, and to be equally at home on grass, stony ground, or rocks. A familiar instance of this fact may be found in our Exmoor ponies. They have ample choice of ground, but of their own will they prefer rough and rocky ground, climbing and leaping with an activity and sure-footedness that is hardly surpassed by the goat itself.

A curious instance of this capacity occurred to a friend of mine, a mighty Nimrod, who has hunted in many parts of Europe, Asia, and Africa.

He was stationed with his regiment in India, and was fond of going out alone to hunt, mounted on a little Circassian horse. On one of those excursions he lost his way, and, finding himself at the foot of a rocky cliff, determined to ascend it so as to find his bearings.

So he dismounted, and began to climb up the rock, leaving the horse with the bridle over its neck. These horses are trained to stand still while the hunter goes off on foot in search of his game. When Col.

D—— had mounted about half-way up the precipice, he heard a scrambling sound beneath him, and on looking down saw his horse in the act of following him.

The rock was very steep, but the horse could climb it as well as the man. There was one part of the ascent of which Col. D—— sent me a sketch. In one place the rock projected into a sort of buttress nearly smooth and perpendicular, and it was necessary to work round it before reaching a firm foothold. A narrow ledge ran round it, just wide enough for the toes to rest upon, and, by means of clutching with his fingers at any irregularity of the face of the rock, the hunter managed to work his way round the obstacle.

To his great astonishment, he saw that the horse was following the same track as himself, and had managed to sidle round the buttress in exactly the same manner. The horse reared itself upright, set the toes of the hind feet upon the narrow ledge, clung to the rock with the sharp-edged toes of the fore feet, and so contrived to achieve the dangerous passage. Of course the animal was unshod.

An admirable example of climbing power possessed by the unshod horse is given by Lady Florence Dixie, in her work 'Across Patagonia.'

'Soon our horses began to neigh and prick up

their ears as we advanced beyond a clearing. Their cries were answered from somewhere beyond us, and, pushing forward into the open, we came upon a herd of wild horses, who, on hearing our advance, had stopped grazing, and now they stood collected in a knot together, snorting and stamping, and staring at us in evident amazement.

'One of their number came boldly trotting out to meet us, and evidently with no pacific intentions; his wicked eye and his white teeth, which he had bared fiercely, looking by no means encouraging. But suddenly he stopped short, looked at us for a moment, and then, with a wild snort, dashed madly away, followed by the whole herd.

'They disappeared like lightning over the brow of a deep ravine, to emerge again on our view after a couple of seconds, scampering like goats up its opposite side, which rose almost perpendicular to a height of six or seven hundred feet. They reached its crest at full gallop in the twinkling of an eye, and, without pausing for an instant, disappeared again, leaving us wondering and amazed at their marvellous agility.

'I had often seen their paths leading up hill-sides which a man could scarcely climb, but, till now that I had witnessed a specimen of their powers with my own eyes, I had scarcely been able to believe them possessed of a nimbleness and cleverness of foot which would not discredit a chamois.'

HERE are some of the offices which have to be performed by the horse's hoof, together with the reasons for those offices.

The horse is a very heavy animal, and therefore the hoof which carries that weight must be STRONG.

Strength might be obtained by solidity, but solidity would involve weight, so that the horse would have to lift several ounces each time that he made a step. This may not seem very much in detail, but in the aggregate it is very considerable indeed.

Moreover, the power which is required to raise the foot is very much in excess of the weight to be lifted. The hoof is at the extremity of a long lever, the power of which is applied at the end of the shorter arm, so that, as has been roughly calculated, an ounce on the hoof is equivalent to a pound on the back. The horse is not furnished with muscles fitted for lifting heavy weights at the ends of its legs, and therefore the foot is carefully made as light as possible. The hoof, therefore, must be LIGHT.

It must be HARD, so as to endure contact with sharp-edged rock, a sun-baked soil, or loose stones.

It must be ELASTIC, in order to obviate the jar which would be caused by the concussion of a hard and unyielding substance with the hard and unyielding ground.

It must be SHARP-EDGED, to give the animal a footing on rocks or uneven ground.

It must be CLINGING, so as to save the horse from falling on a wet, slippery, or frozen surface.

Lastly, as the hoof must be perpetually worn away, it must be capable of SELF REPAIR in exact proportion to the loss of material. All these apparently conflicting characteristics are to be found in the hoof of the horse in its natural state, and there is not one of them which man does not impair, or actually annul, in his attempts to improve upon nature.

The provisions for combining lightness with strength begin with the bones. In order to support the large hoof, it is necessary that the terminal, or 'coffin' bone should be enlarged. If, however, this bone were solid, several ounces would be added to the weight. It is, therefore, made of a 'cancellous' structure—*i.e.* somewhat sponge-like in appearance. This formation is evident, even from the outside, but is very much better shown by making a section of the bone. A somewhat similar structure may be seen in the skull of the elephant, where size has to be conjoined with lightness.

There is now before me the coffin bone of a carthorse. It measures nine and a half inches round the edge, four and a quarter inches across, from wing to wing, and only just exceeds four ounces in weight.

A similar principle is carried out in the hoof, but, as the material is horn and not bone, it must be treated in a different manner.

In the hoof there are three distinct kinds of horn. Each kind is secreted from a different source, and is perpetually renewed when it becomes worn out or effete. The three horns are, firstly, the outer 'CRUST' or 'WALL;' secondly, the 'FROG,' which occupies the centre of the under surface of the hoof, and ought to bear the weight of the horse when it first sets foot to the ground; and, thirdly, the 'SOLE,' which connects the frog with the wall.

We will take each of these structures separately.

Even the WALL is not a solid piece of horn, but is made of a great number of very thin horny plates called 'laminæ.' These laminæ are shaped like knife-blades, the backs being very much thicker than the edges. If we can imagine some six or seven hundred of these horny blades to be set closely side by side, their backs to be fused together and their edges free, we may form some conception of this portion of the hoof. They have been well compared to the gills of a mushroom.

These laminæ are, in fact, flattened hairs, and, like all hairs, grow from their roots, which may be found in the '·coronary ring' which surrounds the upper portion of the hoof. The laminæ which form the front of the hoof grow much more rapidly than the others, so as to compensate for the greater wear and tear which takes place on the toe than on the ' heel,' as it is called.

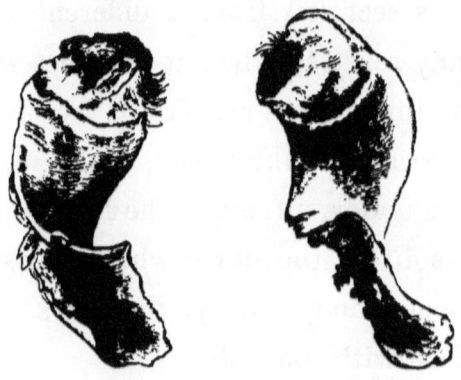

HOOF OF FALKLAND ISLAND HORSE
(FRONT AND BACK VIEWS).

Therefore, unless the front of the hoof be regularly worn away, as was intended by the Creator, or cut away by the knife, according to man's ideas of improving on nature, it grows to an abnormal length, and prevents the animal from walking in a natural manner.

The accompanying illustration represents the overgrown hoof of a horse which had lived on the soft and swampy ground of the Falkland Isles, and which was therefore unable to wear away the hoof by friction. As may be seen from the illustration, the hoof had curled round like a ram's horn, forcing the animal to walk on the side of the hoof and not on its face. As to the frog and sole, it is not easy to identify them, so completely are they enveloped by the overgrown horn of the wall.

The accompanying illustration is drawn from a specimen in the Museum of the College of Surgeons. It may be seen among the series of epidermal appendages, and forms a good reply to those persons who think that the hoof was formed for marshy ground.

Should the reader be a keeper of cage birds, especially canaries, he will remember that both the claws and beak are apt to be overgrown for want of the friction to which they would be subjected had the bird lived a natural life. Unless both beak and claws be artificially cut, as a succedaneum for the natural friction, the bird will be greatly inconvenienced, and may probably die from inability to procure food.

A similar result of overgrowth for lack of friction is often seen in rabbits, mice, rats, and other rodent animals. One of the incisor teeth is accidentally broken, and the corresponding tooth of the opposite jaw not having anything to rub against, continues to grow until the tooth attains an astonishing length.

Those of the lower jaw will sometimes curl over the top of the head, while those of the upper jaw, on account of the greater curvature of the socket, will continue their growth until they form a perfect ring. Examples of these overgrowths may be seen in the College of Surgeons.

Rabbit-keepers must have noticed how their pets are continually moving their jaws as if masticating something. This movement so closely resembles that of ruminant animals when they chew the cud, that the Jews of old times took for granted that the hare and the cony—*i.e.* the hyrax—chew the cud, whereas they only rub the upper and lower teeth against each other for the purpose of preserving the chisel-like edge.

ANOTHER needful characteristic of the hoof now comes before us. In order to avoid jarring the brain and spinal cord at each step, it is necessary that the hoof should be ELASTIC.

As I have already mentioned, the general elasticity of the whole frame is largely due to the mode in which the joints of the limbs are made. The peculiar angle at which the fetlock is set has also an influence on the elasticity, and horsemen are well aware that when the pastern is too upright they feel jarred at every step.

I always endeavour to show how man's inventions, especially in practical mechanics, have their prototypes in nature. If the reader will compare the accompanying illustration with the fetlock of the horse on page 15, he will see that the angle of the connecting piece (marked $a'$) is almost identical with that of the fetlock.

But something more than the mere arrangement of the bones is required. The portion of the hoof which comes first to the ground at each step must be elastic, and this requirement brings us to one of

SIDE BEARING OF RAILWAY CARRIAGE.

the most important portions of the horse's hoof, it being the chief source of elasticity. This is the FROG, so called, because when untouched by the knife it really bears some resemblance to a crouching frog.

It occupies the centre of the lower surface of the hoof, and, as the reader may see from the accompanying illustration, looks something like the letter V. The rounded portions at the ends of the bars are called the 'glumes,' or heels of the frog.

When the hoof is left in its natural form, the frog fills up a considerable portion of the hoof. It is not hard like the wall, which, if properly treated, becomes so hard that a knife will scarcely touch it, but is quite soft and elastic, feeling when handled much like vulcanised indiarubber.

As the horse steps, the weight first comes on the hinder, or heel portion of the frog, then upon its centre, and afterwards upon the wall.

Two objects are fulfilled by this structure. Firstly, by means of this elastic material interposed between the horse and the ground, the animal treads softly, and does not 'jar' the body, as would be the case if the bearing came first on the wall. Horses are instinctively aware of this fact, and when at liberty in a field they may be seen shuffling about in order to obtain the central bearing, for which the hoof was made.

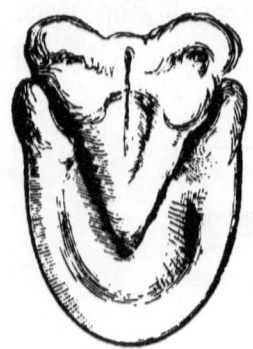

UNSHOD HOOF OF WILD HORSE.

Again, man has imitated nature in artificial locomotion.

In many forms of carriages, especially those of railways, the weight rests upon the central portion of the springs, as seen in the accompanying illustration. The inventors of this form of spring, and of the diagonal attachment which is figured on page 33, were probably unaware that their inventions had been anticipated in nature—not only for many years before carriages were invented, but many ages before man could have existed.

The next object is to prevent the horse from slipping on wet or smooth surfaces. For this object the frog is wonderfully adapted, as it will cling to ice or a wet boulder, and enable the animal to traverse a slippery surface with perfect freedom.

When, therefore, the hoof of an unshod horse comes to the ground, and the weight of the animal rests upon it, the hoof has a double hold, the frog in

CENTRAL BEARING OF RAILWAY CARRIAGE.

the centre clinging like indiarubber, and the sharp-edged wall holding to the least roughness or irregularity.

Like the wall, the frog is subject to perpetual wear, and therefore must be perpetually renewed. It is not, however, rubbed down by friction, as is the wall, but its outer portions continually become detached in little loose flaps, which hang on for a time and then break away altogether, so as to make way for the fresh material which has been formed above them.

Lastly, there comes the SOLE, which binds the frog and the wall together. The horn of which this portion of the hoof is made is very different from that of the wall or the frog. It is formed of a number of extremely hard and strong horny plates laid one above the other, and curved so as to form a sort of dome surrounding both sides and the front of the frog. The sole has another object beside connecting the frog and the wall. It is intended to defend the sensitive parts of the interior hoof from stones, sharp points of rocks, and so forth. When the sole becomes worn out, it has the faculty of reproducing itself in a manner quite distinct from that of the wall and the frog. Instead of being rubbed away by friction like the former, or throwing off little flaps like the latter, it exfoliates in flakes, a new flake being secreted above before the effete one falls below.

One point more about the hoof remains to be mentioned.

As the horse is intended by nature not only to go on level ground but to be able to climb rocks, it is necessary that the edges of the hoofs should be sharp and the interior slightly concave. It must be evident, therefore, that if the edges be blunted and flattened, or, still worse, if they be rounded, especially at the toes, one function of the hoof cannot be exercised.

The reader will now see that these three kinds of horn work with each other, and that if either of them be removed the other two cannot perform their tasks at all, and that if one be even injured, the others are proportionately weakened. Each of these points will presently be considered at length, but I have thought it better to begin by giving the reader a general idea of the natural hoof and its structure. At all events, enough has been said to show that the hoof is not made for one kind of ground or one climate only, but is capable of sustaining the animal on rock, marsh, loose stone, or ice.

We will now take a corresponding view of the internal structure of the hoof.

## CHAPTER III.

Internal structure of the hoof—The hoof compared with the Jacquard loom—Longitudinal section of pastern—The 'navicular' bone: its origin, form, and office—The tendons—Flexor and extensor muscles and their tendons—The 'coronary ring' and its object—The hoof of the horse and the nail of man—The 'quick' of the nail—The 'sensitive' or 'vascular' laminæ—Mutual dependence of the horny and vascular laminæ—Analogy of laminæ of whalebone and those of hoof—Expansive property of hoof—Mr. Miles's experiments—Advantage of this property in a hunter—Effect of shoes in leaping—In any pace the heel comes on the ground before the toe—Importance of this fact.

THROUGHOUT the whole animal kingdom there is, perhaps, no structure which is more elaborately intricate than that of the internal hoof.

Yet, intricate as it may be, it is based on one leading idea, so that all the multitudinous details subserve one single purpose. The same principle is observed in many of the complicated machines invented by man. Take, for example, the various modifications of the Jacquard loom, especially that form which produces ribbons into which various patterns are woven. To the novice, nothing looks more hopelessly elaborate and confusing, while to the skilled weaver nothing can be simpler or more intelligible.

There are hundreds of threads crossing and recrossing each other in apparently hopeless confusion. There are several rows of little shuttles, each fitted with differently coloured silk. There are slender rods and cranks, and over all this mixture of brass, steel, strings, silks, and shuttles, there is a string of cards perforated with circular holes, each turning over as the shuttles dart from one side of the machine to the other.

If these parts be separated, they seem to have no connection with each other. A person who did not understand the principles of the machine might think that you were joking if you showed him the string of perforated cards and a patterned ribbon, and told him that the holes on the cards constituted the original pattern, which was reproduced in a different form on the silk, and that a skilled weaver can read off one from the other as a linguist can read English into Greek, or *vice versâ*. Yet all these multitudinous details are arranged by one master mind, and all work harmoniously together to one single end.

So it is with the horse's hoof.

That the three kinds of horn of which the external hoof is constructed should be formed so as to act in concert with the infinitely more elaborate internal hoof, seems at first sight as incredible as that the holes in the cards above the loom contain the pattern which the machine works in coloured silks below.

Yet, as I have said, the manager of the machine can read the pattern of the silk in the cards, and he knows that if only a single card were to be removed, or even transposed, or a hole omitted, the pattern

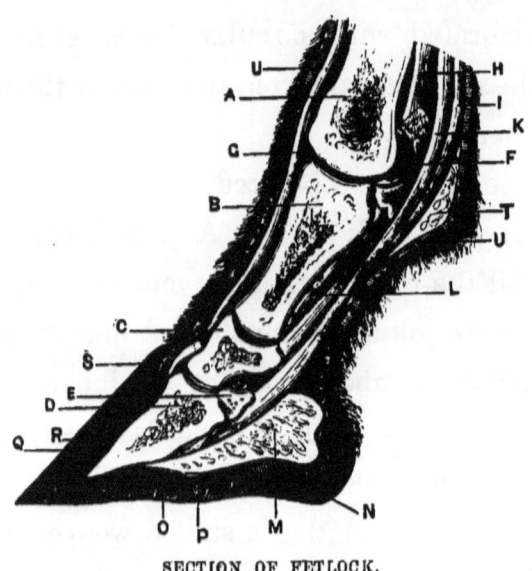

SECTION OF FETLOCK.

would be imperfect. Similarly, those who know the construction of the hoof are perfectly aware that all its parts, whether external or internal, are dependent on each other, and that an injury done to one will affect all the others.

I shall now endeavour to place before the reader the interior of the hoof, and its connection with the horny covering that surrounds it, and the limb of which it forms the extremity.

The accompanying illustrations represent a longitudinal section of the lower portion of the horse's foot.

Beginning as before with the bones, A is the lower part of the cannon bone, or metacarpal; B is the Long Pastern, C the Short Pastern, and D the Coffin Bone. Another bone, shown at E, now comes before

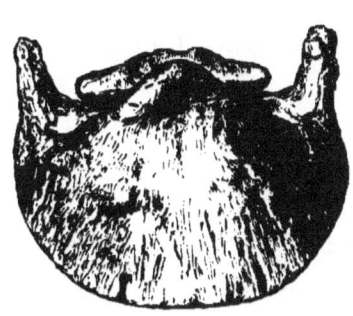 

FRONT VIEW OF COFFIN AND NAVICULAR BONES.   BACK VIEW OF COFFIN, NAVICULAR, AND SHORT-PASTERN BONES.

us. It is quite a little one, but is of the greatest importance to the hoof. Its scientific name is the 'Navicular' bone. The word is Latin, signifying a little ship or boat, and is given to the bone because it somewhat resembles a birch-bark canoe in shape. The too-familiar 'navicular' disease originates in the structures surrounding this bone. Another name is the 'shuttle-bone.'

As the section on page 40 does not give any idea

of the true shape of either the coffin or navicular bones, two views of each are here presented.

What might be the origin of this bone was for some time a matter of controversy. It was long thought to be a modification of one of the missing bones of the foot, but is now known to be a 'sesamoid' bone. There are many of these bones, variously formed in different animals, the largest being the knee-cap, or 'patella.' They are developed within the tendons, and derive their rather fanciful name from their usually small size, together with their shape, which bears some resemblance to that of the sesame seed.

The reader will see that the navicular lies behind the coffin bone, and between the two wings. Its chief object is to act as a pulley, so as to enable the foot to be bent, or 'flexed,' with greater force.

Another sesamoid bone, performing a similar office, may be seen at F, just at the end of the cannon bone, and aids in producing the characteristic projection at the beginning of the pastern. At T is seen the fatty cushion of the pastern. Its structure acts as a guard to the delicate machinery of the sesamoid bone and tendon.

We are now led to another element in the horse's foot — namely, the tendons — *i.e.* the ligamentary bands by which the muscles act on their attachments.

Whenever a limb or any portion of the body has to be bent, there are always two opposite sets of muscles, one called the 'flexors,' because they bend the limb, and the other 'extensors,' because they straighten it. As a rule, the former are far more powerful than the latter.

This we may see exemplified in the well-known schoolboy experiment. When one boy presses together the tips of the forefingers, and another grasps his wrists, and tries to pull the fingers apart at right angles to the body, the latter cannot succeed unless he be very much bigger and more powerful than the former. The reason of this discrepancy is, that the first boy uses his flexor muscles, while the second boy employs the extensors. A very familiar example of a flexor muscle is to be found in the 'biceps,' of which an athlete is apt to be exceedingly proud.

The tendons, which in fact surround the muscles as a thin membrane, and are brought to a cord-like shape at the extremities, take their name from their muscles, and of course are of proportionate strength. I may here mention, as another very familiar example, that the 'stringiness' of meat is due to the thin coating of tendon which envelopes the muscles.

At I and K is seen the double flexor tendon. This ligament divides after passing over the sesamoid bone of the fetlock, the front portion being attached to the upper part of the short pastern bone, and the other to the floor of the coffin bone, after passing over the navicular bone. This tendon, therefore, twice changes its line, each time gaining an increase of power.

In the front of the limb runs the extensor tendon. It also divides into two portions, one being attached to the short pastern bone and the other to the coffin bone. These two tendons, therefore, or rather the muscles to which they are attached, would continually pull against each other were it not for the provision that the one is always relaxed in proportion as the other contracts. At II is shown the suspensory ligament, the name of which indicates its use.

These tendons and their attachment can readily be observed in a dried specimen, although it is more satisfactory to study them from the recent subject. I have now before me a section of a pastern which was made many years ago, and has been subjected to very hard usage. It is, however, in perfect preservation, and, unless wilfully damaged, will remain unchanged for centuries.

I recommend the reader to colour all these anatomical illustrations, every one of which has been drawn from the actual object.

At s, on page 40, is shown the coronary ring, which secretes the horn of the wall. Being extremely vascular —*i.e.* filled with arteries and veins, it shrivels almost to nothing after the death of the animal, and in a dried specimen there is only a space to show where the coronary ring has been.

At M is shown the soft frog, which is guarded by the hard or exterior frog, which is seen at N. This, however, is not quite large enough. The draughtsman, who, I regret to say, died almost suddenly before the series of drawings was completed, could not obtain access to a specimen which had not been mutilated by man, and therefore had to draw the frog as he saw it.

P shows the internal or sensitive sole, and O the hard or external sole, which has already been mentioned. R shows the wall or crust, and U is the skin.

HORNY LAMINÆ.

Now we will see how the sensitive structures are connected with the external hoof, which, it must be remembered, is the analogue of the human nail.

The reader will remember that on page 29 the six hundred laminæ of which the wall is composed were compared to knife-blades, the edges being inwards. Upon the coffin bone are set edgewise a corresponding number of vascular, or sensitive laminæ, shaped very much like the horny laminæ, and inserted between them like partially interlaced fingers. They hold the hoof so tightly that to pull it off, even when the pastern is separated from the rest of the limb, is exceedingly difficult.

In the accompanying illustration is represented a section of a hoof which was made for me at the College of Surgeons. E is the bone. At A B are the horny and sensitive laminæ interlacing with each other, the latter being as amply supplied with nerves as the base and 'quick' of the human nail.

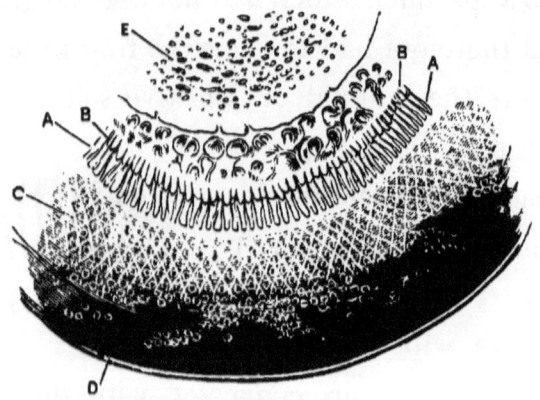

SECTION OF HOOF THROUGH THE COFFIN BONE.

C is sensitive structure, and D is the external horn.

This figure affords an excellent example of the curious alteration in appearance which is made by a section. Even in the longitudinal section of the pastern, as shown in p. 40, the coffin bone, owing to the absence of the wing, looks scarcely half its real size, while no one could form the least idea of either the size or shape of the navicular bone from the section.

There is now before me a dried specimen of a pastern, from which the hoof has been removed, the sensitive laminæ having been allowed to remain in their places.

They have, of course, become shrivelled out of all shape, have lost their bright scarlet colour, and round the edge of the coffin bone they are quite undistinguishable, the bone looking as if it were enveloped in a rough brown skin. But at the upper part of the bone, especially just below the coronary ring, the separate laminæ can easily be detected, so that when the hoof is shown beside them the connection is perfectly intelligible.

I have already mentioned that there are about six hundred horny laminæ in each hoof. If we make a careful section of the laminæ, we find that, slight and delicate as they are, their structure is far from being simple. Hard at their bases, which form the wall of the hoof, they become more and more fragile towards their edges, so that the portions which interlace with the sensitive laminæ might be rubbed to pieces between the finger and thumb.

In order to be seen in their full perfection, a very thin section of them should be made, and then be viewed by polarised light. A very moderate power, say a half-inch object-glass, is required. They will then be seen to have their surfaces covered with sub-laminæ, while now and then one of them throws off a branch similarly furnished with sub-laminæ. The colours which are presented by these structures under polarised light are beyond the power of

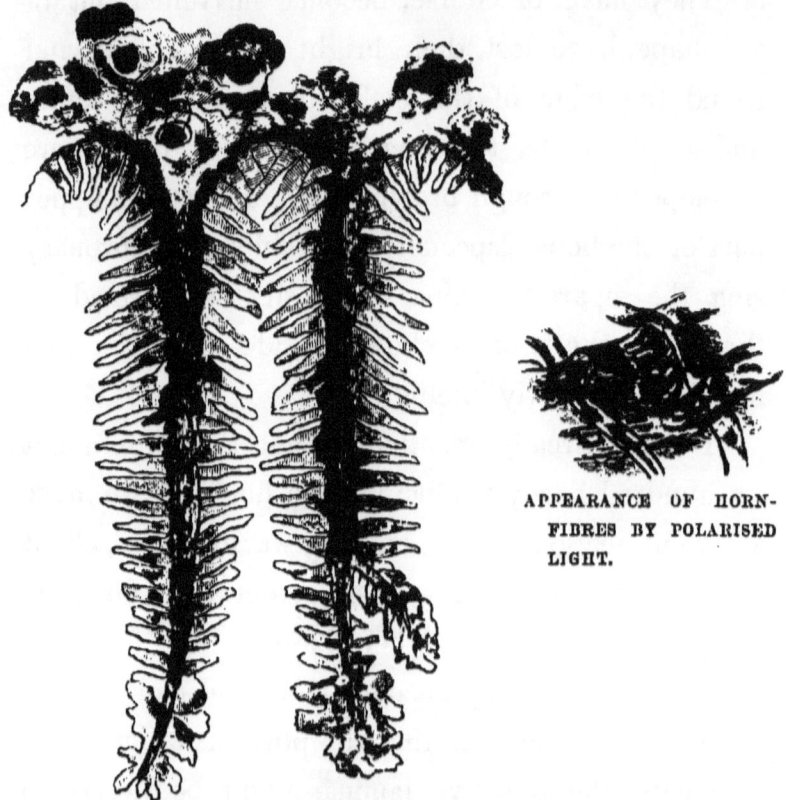

APPEARANCE OF HORN-FIBRES BY POLARISED LIGHT.

TWO LAMINÆ WITH THEIR SUB-LAMINÆ.

description. I have, however, endeavoured to give some idea of them without the aid of colour.

One other point remains to be mentioned in connection with this part of the subject. This is the 'linea alba,' or white line, which marks the junction of the sole and wall. It is analogous to the 'quick' of the human nail, and is exceedingly sensitive.

In the untouched hoof, this line, which is as vulnerable as the heel of Achilles, is thoroughly protected by the thick and strong horny layers of the sole; but if that protection be removed, or even weakened, the pressure of the horse's weight against a stone or similar object will cause such intolerable agony, that the animal will drop as if shot when he treads upon it.

Now we will see how all these structures, which are apparently so different, can work together in harmony.

Suppose the animal to be walking. At each step a considerable part of the weight of the horse is thrown upon the hoof. The first portion to come to the ground is the elastic frog, and as the frog yields to pressure, the sharp-edged wall also comes upon the ground, so that the horse is partly supported by the frog and partly by the wall.

The pressure of the frog is transmitted to the sole, and thence to the wall, which slightly expands. There has been great controversy about this expansion. Some writers, judging apparently from the dried hoof, have denied that any expansion at all takes place. But it must be remembered that the hoof, when it belongs to the living animal, is a very different instrument from the same hoof when dry. In the latter state it is hard and unyielding, but in the former, though hard on the exterior, it becomes gradually softer in the interior, and is perfectly capable of yielding to pressure.

Similarly, 'whalebone,' as we popularly call it, or 'baleen,' as it ought to be termed, is a kind of horny structure, formed, like the hoof, of a vast number of hair-like fibres fused together. The whalebone of commerce is hard, and though very elastic to a certain extent, is sure to snap if bent too far. This is well exemplified by the pictorial advertisements which meet, or rather which are forced on our eyes at every turn, and which represent ladies in despair about their corsets, the 'bones' of which *will* break.

But the baleen, as it exists in the living animal, is a very different structure. It is many feet in length, very nearly straight, as soft and pliable as a fresh tendon, and quite as incapable of being broken when bent. Indeed, it is so pliable, that when the whale closes its mouth, the ends of the baleen slip into a deep groove on either side of the lower jaw, the long plates, or laminæ, bending nearly at right angles. They remain there until the animal opens its mouth, when the baleen springs back again into its previous form.

So it is with two out of the three kinds of horn which constitute the hoof. The frog yields vertically and the wall laterally, the rigid sole serving to transfer the pressure as has already been mentioned. Consequently, as long as the animal moves, the horn is never at rest. It is perpetually fulfilling the tasks for which it was made, is continually thrown off, and as continually replaced. Any interference with nature, therefore, must of necessity be injurious to the hoof.

The expansion is not nearly so much as some of its advocates think, but still, there is quite enough to keep the fibres of the three horns in motion. Mr. G. Ransom ('Free Lance'), in his 'Horses and Roads,' mentions a case where the amount of expansion was carefully measured. The experiments were made by Mr. Miles, a Devonshire squire, author of 'Miles on the Horse's Foot.'

'The subject of them was a horse nine years old, which had always worn shoes since he was first put to work, and had the shoes removed on purpose for the investigation and experiment.

'The unshod foot was then lifted up, and its contour traced with the greatest precision on a piece of board covered with paper. A similar board was then laid on the ground, the same foot was then placed upon it, and the opposite foot held up whilst it was again traced.

'The result was, that it had expanded one eighth part of an inch at the heels and quarters; and from the quarters towards the toe this gradually diminished, showing a space of four inches front, and two inches on each side of the centre of the toe, where no expansion whatever had taken place; the tracings proving at the same time that expansion was only *lateral*, and that none took place in the length of the foot from heel to toe.

'He states that he had other horses which had before shown a still greater expansion than this; but this was only whilst the horse was standing, and upon three legs.'

The expansive property of the hoof has another advantage. When an ordinary horse—say in hunting—is traversing stiff and clayey ground, its progress is greatly hindered by the soil which adheres to the hoofs, and by the depth to which the foot sinks in the ground. Sometimes, when the horse has taken a leap, the fore-hoofs sink so deeply that the animal cannot instantly extricate them. The forward impetus cannot be checked, and the result is that the leg is broken at the pastern, and the horse has to be killed. The rider may probably be killed too, but we are now dealing with the horse and not with its rider.

But, with an unshod horse, the result of a similar leap would be very different. As it comes to the ground, the hoof expands, and consequently gives a wider bearing, so that the foot does not sink so far, and necessarily makes a larger hole. Then, as the hoof contracts as soon as the weight is taken off, it comes easily out of the hole, and so there is but little risk of snapping the pastern.

Even on hard ground, the disadvantage of interference with nature is equally shown. Mr. Miles, who has already been mentioned, writes as follows:—

'When a hunter is shod in the usual manner with seven or eight nails, some are always, for the sake of security, placed in the inner quarter, which is the most expansive portion of the hoof. Let a horse so circumstanced be called upon to leap from a high bank into a hard road, and what happens?

'The weight of the horse and its rider is thrown with an impetus, which greatly increases that of both, upon the bones of the feet, both sides of which are so fettered that neither can yield to make room for them. Consequently, they squeeze the exquisitely sensitive lining of the hoof between their own hard substance, the unyielding horn, and the shanks of one, two, or three nails, as the case may be.'

That the heel comes first to the ground, is followed by the frog, and that the toe only comes on the ground when the horse is standing still, or when it lifts the foot from the ground, has been conclusively proved by the well-known veterinary surgeon, Mr. J. I. Lupton.

'The foot of a living horse in a state of rest remains firmly on the ground—that is, the toe and the heel are on the ground at one and the same time. But if, during this position, the extensor muscles were to contract, then the toe would be raised from the ground. If, on the other hand, the flexor muscles were to contract, then the heel would be raised from the ground.

'Now, during progression, the first movement which takes place is the contraction of the flexor muscles, by which (together with the muscles of the arm) the foot is raised, the toe being the last part of that organ raised from the ground. The foot is now in a position to be sent forward, which is brought about by the contraction of the extensor muscles. The foot is then thrown out as far as the flexor muscles will admit, and, when at the greatest allowable point of extension, the heel is brought in a position with the ground.

'The flexors now in their turn contract, the heel is first raised from the ground, and *lastly* the toe, which brings me back to the point I started from.

'Three principal impressions are made on the foot during progression, namely:—

'(1) On the heel, when great expansion and

yielding takes place, owing to the pressure on the frog, which is forced upwards, causing the ultimate expansion of the walls of the hoof, &c.

'(2) On the middle part of the foot, when the bones bear the weight of the body, the flexors and extensors being for the instant in a state of quietude —*i.e.* neither of them are extending or contracting.

'(3) In the toe, when the animal gives a push, by which an impetus is given to send the body forward.'

These words were written as long ago as 1858, and have since received exact corroboration in the instantaneous photographs of the horse's action, with which we are now so familiar.

It may seem to some persons a matter of little or no consequence whether the heel or toe should first come to the ground during progression. In reality, it is of the very last importance, and, as we shall see in a future page, does not only affect the hoof, but is intimately connected with the whole of the muscles which the horse uses in progression, and with the nerves which supply these muscles with energy.

The reader is requested to bear in mind that the hoof only expands laterally, and that on and near the toe the expansion is so trifling, that it may be practically disregarded. Reference will again be made to this fact.

## CHAPTER IV.

Expansion acknowledged, acted upon, and perverted—Authorised directions for shoeing—'Thinning' and 'opening'—Horseshoeing in England and America—'Dew-drops,' *i.e.* oozing blood—The diseases called by the name of 'thrush'—True causes of thrush—Supply of blood to hoof—Nature's balance of supply and waste—Office of the blood—'Thrush' a safety-valve for inflamed blood—Derivation of the word 'founder'—'Thrush' never seen in wild horses—A comprehensive challenge—Where do wild horses keep their knives?—The frog again—Lieutenant Douglas on the frog and its value. Mr. T. Gepp's experience.

SOME of those who asserted, and rightly, that the hoof expanded when the weight of the horse rested upon it, attempted to improve nature by art, and, with the very best intention, contrived to do a wonderful amount of damage to the animal. Of course, they assumed that the horse must be shod in the usual manner. Then, they forgot that the hoof would not expand unless the weight of the horse rested primarily on the frog, the 'mainstay of the foot,' as Mr. Douglas calls it; and lastly, most of them forgot that the iron shoe would prevent the hoof from either expanding or contracting. Some, however, did recognise this fact, and invented shoes in two pieces, so that expansion was not hindered.

So, thinking that the thick and rigid sole was an obstacle to expansion instead of being one of the means by which expansion is accomplished, and also assuming that the expansion was in all directions, and much greater than is really the case, they believed that by thinning the interior of the hoof they would increase its power of expansion. So they issued the following directions to the farrier when he prepares a horse for the shoe:—

'Pare the sole until it yields to the pressure of the thumb.

'Cut the walls down until they are but little higher than the contiguous sole, taking care to shorten the toe if necessary, it being frequently left too long.

'Cut away the bars, so as to make a gradual slope from the wall to the bottom of the commissures, which must be deepened.

'Lower and *open* the heels, taking the bearing off them for at least an inch on each side of the frog, so that the walls at these parts will not be in immediate contact with the shoe when first put on.

'Pay special attention to the removal of the " pegs " (the hard horny substance which grows down at each side of the frog, and contiguous to it).

These pegs are apt to contract the foot, or make it thrushy, by pinching and narrowing the frog.

'The frog may be pared to stimulate its growth, and the cleft opened; otherwise it may be left untouched,' &c.

These directions for shoeing on 'improved principles' are taken verbatim from the 'United States' Army Tactics on Horseshoeing,' and are quoted by Mr. M. I. Treacy, Veterinarian to the 7th Cavalry, in an article on Horseshoeing in *The United Service Magazine* for February 1884.

One excuse—I cannot call it a reason—for opening the heel and thinning the sole would be comical, but for the injury which it has wrought to thousands of horses. The advocates of the knife say that they are aware of the sensitive and delicate structure of the interior of the hoof, and that the operations of opening and thinning are necessary to preserve those structures. According to them, in this climate the horny parts of the hoof have too rapid a growth (why or how is not stated), so that they prevent expansion, and pinch the internal structures. So, by 'thinning the sole until it yields to the pressure of the thumb,' and cutting through the heel, they think that the internal structures are relieved from constriction! Mayhew very quaintly and truly says that it would be just as sensible to thin and open the skull for the purpose of protecting the brain.

Here is a figure of a hoof in my possession which has been treated according to these directions. It was drawn by the late Mr. Sherwin from the actual specimen, which was by no means the worst that I have seen.

I will now proceed to show what is the result of thus preparing the hoof, first, however, mentioning that the directions which are here given are followed both in England and America. While on a tour through the Northern States in the winter of 1883, I delivered lectures on this subject in many places. One well-known writer and lecturer took umbrage at my statements, and had the hardihood to assert, both by pen and on the platform, that in America the frogs of the horses were not cut away, nor the soles pared. Yet, being in America, I took my descriptions, not from our English customs but from American sources, one, as the reader may have seen, being stamped with official authority.

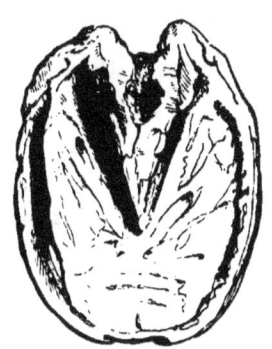

HOOF IMPROVED BY THE FARRIER.

The first direction is, to pare the sole until it yields to the pressure of the thumb. In many cases this paring of the sole --I again use American authorities—is carried so far that blood-specks ('dew-drops,' as they are euphemistically termed) ooze through the thinned horn. Just see what this means. One of the chief duties of the sole is to protect the sensitive structures within the hoof from sharp stones and similar objects.

It is difficult to persuade many people that the Creator really did know how to make a horse, and that divine handiwork cannot very well be improved by man. But, a horse, whose hoof is left as Nature made it, cares nothing about pebbles or even broken flints, but can gallop among them without being even aware of their presence, so dense and strong is the horn of the sole.

Within the last four weeks, an unshod mare belonging to one of my friends ran away with her mistress, who was driving her in an ordinary chaise. The road had been newly laid with loose stones, but she galloped for nearly two miles before she was stopped. I examined 'Dolly's' hoof a few days afterwards, and found that they had suffered no injury from the sharp points and edges of the stones. A shod horse could scarcely have escaped laming.

But, when the sole is pared until it is not so thick as an ordinary visiting card, it is evident that the pressure of a stone must cause the severest pain. Moreover, this paring of the sole exposes the 'linea alba,' which, as I have already mentioned, is identical with the 'quick' of the nail, and equally sensitive.

Unfortunately, when a shod horse picks up a stone, the intruding object is always jammed between the shoe and the sole, and must press upon the linea alba. An unshod horse which has perfect hoofs cannot pick up a stone, for the simple reason that there is no room for it. Let the reader look at the figure of an unshod hoof on p. 34, and he will see that it would be perfectly impossible for the animal to pick up a stone, the frog filling almost the entire cavity of the hoof.

Having now weakened and thinned the sole as much as possible, the farrier is directed to cut down the wall until it is scarcely higher than the sole. This proceeding is ingeniously contrived so as to bring the weakened sole within reach of the smallest pebble.

Next, the bars are attacked, and then the heels 'opened'; this last operation being analogous to removing the keystone of an arch. With regard to the mode in which this system is carried out, Mr. Fleming writes as follows:—'I have seen in forges, where horses were shod on "improved principles," the blood oozing from the sole, which had been pared as thin as parchment, as well as from the frog, through the semi-translucent substance of which, so thin had it been made, the bright red and exceedingly sensitive living membrane beneath could here and there be distinctly seen. Not unfrequently, also, I have noticed blood issuing from the bottom of the deep notches cut off nearly as far as the hairs at the heel.'

Then the ' pegs' are to be removed, on the ground that they contract the frog, and cause ' thrush.' It is impossible to surpass this statement in its extraordinary mixture of ignorance and presumption. If it were true, every wild horse, or every horse, previous to its introduction to the farrier, must be liable to thrush ; whereas, no horse that had never been shod could by any possibility be afflicted with thrush.

What is the ailment which we call thrush, and what is its cause? The presence of the disease is made known by the horribly smelling purulent discharge which is formed in the interior of the hoof, and which at last exudes from it. In bad cases, it can be detected at some distance by the sense of smell alone. The word is a sort of generic term, and is applied very loosely to any disease which causes internal inflammation and consequent discharge of pus.

The disease is wholly owing to the shoe, and is thus caused. Again, at the risk of being tedious,

I must repeat that the structure of the hoof imperatively demands that the wall, frog, and sole shall each be called into play, each be worn away naturally, and each be perpetually reproduced.

Although intended to benefit the horse, the ordinary shoe does it a serious injury in frustrating each of these requirements. As the iron of the shoe is interposed between the wall and the ground, the horn of the wall cannot be worn away, especially in front, where the friction is greatest, and where the horn grows fastest. As the frog is cut away, it cannot take the weight of the horse as it ought to do, and therefore communicates no movement either to the sole or the laminæ of the wall, and so there is none of the incessant expansion and contraction which the hoof requires.

I need hardly explain that all repair of organic waste is effected by means of blood, and that the supply of blood is proportioned to the amount of work to be done by it.

Now, if the skin be stripped off the shank of the horse, the bone will be seen surrounded with a series of very large arteries and veins, intended to carry blood into and from the hoof, so as to supply the continual waste of the horn, the arteries and veins exactly balancing each other. But when that waste is checked, the balance of circulation is destroyed.

The blood is continually forced into the hoof for the purpose of doing its work; but there is very little work to be done, so that the blood cannot be carried out of the hoof by the veins as fast as it is pumped in through the arteries.

Consequently, the smaller vessels become gorged, and the flow of blood impeded. Gradually, congestion comes on, which before long developes into inflammation. Then, the semi stagnant and inflamed blood reaches such a stage that decomposition sets in, and the blood which ought to have passed through the circulation in a natural manner is forced to make its exit in the form of pus, the aperture which it makes for itself being a sort of safety valve not only for the hoof, but for the whole circulatory system.

By the way, the word 'founder,' which is applied to any disease of the sensitive laminæ when it has advanced sufficiently far to lame the horse, is a curious example of the ignorance which seems to be the invariable inheritance of those who have most to do with horses. It is a corruption of the French word *fondre*, signifying to melt, and was used because the farriers believed that the fat of the horse melted out of his body, and ran down his leg into his hoofs!

A corresponding version of the same word is a

familiar term at Oxford, where some of the college servants possess a special recipe for preparing 'fundied' cheese. The cheese is cut up, put into a 'fundy'—*i.e.* a flat vessel made for the purpose, some of the best ale is added to it, and it is heated and stirred until the cheese and beer are melted together to the consistency of treacle.

Thrush has nothing to do with contraction of the frog, though it *is* produced by cutting away the frog, and surrounding the wall with unyielding iron. Yet—probably because thrush shows itself in the neighbourhood of the frog—the opinion that the frog produces thrush is prevalent even among those who ought to know better.

A year or two ago I had been lecturing on the horse's hoof. According to my invariable custom, I began my lecture with a disavowal of any attempt to lay down the law on this very difficult subject. I begged that any of the audience would at once contradict any statement which they thought to be mistaken. I also mentioned that I would wait on the platform after the lecture for half an hour, so that any disputed point might be thoroughly investigated.

No one challenged any remark during the lecture, nor for at least half an hour, during which I remained on the platform after the conclusion of the lecture. But a local veterinary surgeon, who was present at the lecture, wrote a letter to a local journal, saying that I was entirely mistaken in stating that the frog ought to rest on the ground, and asserting, moreover, that thrush was produced by neglecting to pare the frog and cut away the flaps that hang from it.

Anonymous attacks I never answer, but as this man did have the honesty to append his name to his letter, and did not employ personal abuse in lieu of argument, I sent an answer. In it I asked him where wild horses kept the knives wherewith they pared their frogs. Next, I told him that if he could produce a thrushy hoof of a horse which had never been shod, I would first eat my words, and then the hoof, thrush and all.

This event occurred in England, but I have undergone a very similar experience in America, the opponent having lacked the courage to attack me on the platform though he was present at the lecture, and having written letters against me in the local journals, and publicly spoken against me after I had left the country and could not answer.

As we are on the subject of the frog, we may as well continue it.

I have already mentioned that the frog was made for the express purpose of resting upon the ground, so as to perform a threefold office.

In the first place it is soft and yielding, so as to take off the jar which would ensue if the horse trod first upon the wall.

In the next place, it communicates the pressure through the sole to the walls, so as to enable the hoof to expand and contract laterally.

Lieutenant Douglas, in his valuable work on horseshoeing, is very emphatic on this point:—

'A little reflection on this important subject will show how very important frog-pressure is, as, even when the horse is resting as he stands in the stall, some portion of his weight must fall from the lower pastern bone upon the navicular bone, which rests *upon* the back sinew, which in its turn should receive support from the sensitive and insensitive parts of the frog underneath it.

'How much more then must this necessary support be needed when the animal is in motion, especially if at a fast gallop, or landing on hard ground after being leaped over a hedge. If the frog is there to receive the shock, the horse lands on his feet with all the ease and comfort that a cat does upon hers after a jump; but when the frog has been cut away there is nothing to break the fall, and, as is often the case, the animal is ruined by the jar having brought on irritation of the sheath which covers the back sinew, and inflammation sets in.

'La Fosse describes this action as a "compression," fitly comparing the process of the lower pastern bone squeezing the navicular bone on the top of the tendon to the action of the hammer upon the anvil.'

Thirdly, it is intended to cling to smooth and wet surfaces, which its indiarubber-like consistency enables it to do. The late Mr. T. M. Gepp, of Chelmsford, a veteran Nimrod, told me that when he was in Palestine, the horses at first absolutely terrified him by the way in which they sprang from rock to rock, the frog enabling them to cling to any smooth surface on which they might alight, and the sharply edged wall hitching upon the slightest irregularity.

## CHAPTER V.

The FROG continued—'Hammering' on the roads—Cause of the hammering—King Theodore's horse 'Hammel'—Elasticity of the frog—The 'Village Blacksmith'—The bicycle wheel—Analogy between carriage springs and the horse's hoof—Lodgment of stones—Mr. S. Chapin and his bare-footed horse—Hartmann's safety pad—Spiked shoes—American shoeing—Injury to the untouched frog impossible—'Greasy' roads—Manchester 'lurry' horses—The streets of Manchester—A doctor's horses—Shoes of lurry horses—Ludgate Hill and its terrors—Lieut. Douglas's testimony—Indiarubber soles on ice and on board ship—The hoof an organ of touch—Mr. J. Bellows's story—'Free Lance's' view of the frog and its uses—Relative dependence of the frog and navicular bone.

SOFTNESS of tread is an exceedingly valuable property in a horse. We know how 'it ain't the 'untin' as 'urts the 'orses, but the 'ammer, 'ammer, 'ammer, on the 'ard 'igh roads.' The worthy and afflicted groom was perfectly right. This incessant hammering, which is exceedingly injurious to the horse, is entirely owing to the shoe, an unshod horse treading almost as noiselessly as an elephant does, and being delightfully easy to the rider.

The first adult working horse which I ever saw unshod was 'Hammel,' the war-horse of the late

King Theodore of Abyssinia. He was shown in a menagerie, and after I had inspected his hoofs carefully, I remarked to the exhibitor that I presumed him to be kept for show and not for use. But the proprietor assured me that he took his share of the work with the other horses, and that when travelling from one place to another he was regularly harnessed to the vans or carts.

Farriers and grooms are, as a rule, impressed with an idea that because the frog is soft, it will be hurt by coming upon a hard road. So, with the very best intentions, they cut it off either partially or entirely.

There is now before me an entire frog which was cut off at a single stroke of the farrier's knife. It was taken from the floor of a forge by General Carter, a well-known lover of horses. It seems strange that such an idea should prevail in these 'cycling' days, when indiarubber tires have been practically found not only to be more pleasant for the rider than the old steel-faced tires, but experience has proved that the soft elastic indiarubber tire is far more lasting than one which is made of steel.

Suppose that we revert to the parallel of the horse's hoof and the railway carriage spring. In the accompanying illustration, fig. 1 represents the spring, which is made of a number of flat strips of steel, the bearing being at A. Now, supposing that a village blacksmith, who knew nothing about machinery, were to be placed in charge of the engine, he would probably be dissatisfied with the spring as being inadequate to its work, and therefore dangerous. 'It will never answer,' he might say, ' to trust the weight of a heavy engine on a few strips of steel, any one of which might break, upset the engine, and throw it off the line.' So he would cut away the

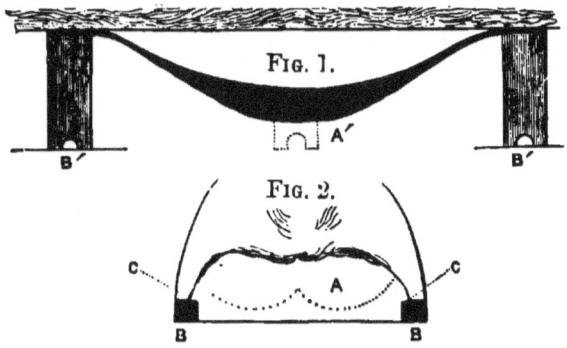

RAILWAY SPRING AND HORSE'S HOOF.

central bearing, as shown by the dotted line at A, and substitute heavy blocks of iron, as shown at B B, thus abrogating the springs, and throwing the bearing to the sides instead of the centre.

This proceeding may seem very absurd, but it is not one whit less ridiculous than the mode in which the farrier of the present time treats the hoof of the horse, and, in fact, is almost identical with it.

In fig. 2 a diagrammatic section of the hoof and

shoe is given, as seen from the front. A represents the double frog, which was intended by nature to rest upon the ground, and form a central bearing like that of the wheel at A in fig. 1. This is more or less cut away, at all events sufficiently to keep it off the ground. Then a thick unyielding iron shoe is nailed on the hoof, so as to take the bearing from the centre, where it was meant to be, and to throw it to the circumference, as at B B, where it has no business, and where it positively injures the animal by causing a jar at every step.

I may here mention that the space which is left between the shoe and the wall, as shown at C C, is the place where stones lodge, and into which, as we know, they become wedged so firmly, that even the 'picker,' with which all horse riders or drivers ought to be furnished, cannot remove it without the use of considerable force. The reader will see how a stone which is thus lodged must press upon the denuded linea alba, and why it is that a horse falls suddenly when such a misfortune happens.

If the reader will refer to the natural hoof shown on p. 34, he will see that no stone can possibly lodge in it.

Lastly, we will take the third duty of the frog— *i.e.* the power of clinging to slippery and smooth surfaces. A hoof which has been untouched by man will scarcely ever slip. This is not theory, but a proved fact, and here is a case in point.

Mr. S. Chapin, of Lowell, Mass., U.S.A., is well known for his love of horses, and for his practical experience with them. Among his horses was one which he had owned about seven years, and from which, after considerable qualms, and against universal advice, he removed the shoes altogether. After some experience he wrote as follows, dating his letter December 15, 1883 :—

'I now drive my horse up hill and down, and over pavements, crossings, &c. I never expect to see a harder winter for ice than we had here in Lowell last year, when some of my neighbours sharpened their horses' shoes twice a week, and I drove all the winter (and pretty sharp too) without a shoe, and without slipping either.' I have before me a long list of English gentlemen who have much the same experience.

Here is a curious example of the involuntary homage done by man to nature.

About four years ago, in the course of a conversation with the late Mr. F. Fordham Flower, whose successful crusade against the bearing rein is well known, the question of 'roughing' horses' shoes in winter was mentioned. Mr. Flower said that there was a very much better plan than roughing—namely, the adoption of Hartmann's Safety Pad, which had the advantage of being easily removed when the horse entered the stable after his work.

He also told me that, having heard of this pad, he took his carriage to Messrs. Martingale, who supply it, and had it fitted to his horses at the door. The animals went so easily, although they had previously been slipping at every step, that he took them to a spot where ice had formed upon the road. The horses went over it at a trot, and from that time he had no trouble with them.

HARTMANN'S SAFETY PAD.

I at once went to Messrs. Martingale, who kindly gave me one or two specimens, and explained the mode of use.

As the reader may see, the pad is composed of thick indiarubber, with three thin steel flanges. By means of a sort of tongs or pincers the pad can be bent longitudinally, so as to allow the flanges to be slipped under the shoe. When the pressure of the tongs is relaxed, the pad resumes its shape by its own elasticity, and is firmly held in its place by the flanges. When the horse comes home, the pad, which is, in fact, an artificial frog, can be removed by the tongs as easily as it was applied.

In the illustration the upper face of the pad is given, showing the groove into which is received the frog which has been cut according to the modern 'improved' system. The lower face, which comes upon the ground, has no definite groove, but is fashioned according to the shape of the untouched frog.

So we have again an acknowledgment that nature is right and man wrong.

NATURALLY, the horse's hoof is furnished with a soft elastic pad, which prevents the animal from slipping. So, man cuts away the natural pad, and because the horse slips, is obliged to furnish the iron shoe with spikes.

Now, however well these spikes are made, no matter whether they be part of the shoe or whether they be added to it, they are soon worn down and become useless. In any case they are contrary to the natural step of the horse, and are awkward for walking. If they be part of the shoe, they cause the greatest inconvenience while the horse is in the stable, and if they be screwed or driven into holes made in the shoe, they are apt to snap off when any great strain is thrown on them.

The result of all this experience is that the Hartmann's pad was invented. It is liable to neither of these defects, is very lasting, but is, after all, nothing more or less than an artificial reproduction of the natural pad which the farrier has cut away in unthinking obedience to routine. As to expense, this pad does not cost much, and when it is worn out, another can be procured at a small price. But the natural pad, or frog, which is necessarily far superior to any imitation, costs nothing at all, and never can be worn out, because it has the faculty of reproducing itself as fast as it is worn away.

I noticed that in America the fact that the frog ought to come to the ground is more generally recognised than is the case in England. But even there the shoe is, as a rule, nailed *upon* the wall, so that the frog has to grow to an abnormal extent before it can reach the ground, and therefore does not have fair play. The only shoe which really recognises the duties of the frog, and does not interfere with them, is the 'Charlier,' which will presently be described.

As for any fear that the frog may be injured by contact with the ground, it may be dismissed from our minds. As was well pointed out by 'Kangaroo,' in *The Field* newspaper, it is impossible for a horse to become footsore in the frog, sole, or heel of its foot, as a result of travelling barefoot.'

I have already mentioned the effect of the safety pad, and *a fortiori* of the frog upon ice. But icy roads are only to be found in winter time, and that not in every year. I believe that in the winter of 1883-4 there was no ice in the streets of London, England seeming to have sent her share of ice and snow to America, where I was staying from October 1883 to April 1884.

Still, many of our roads present surfaces which, under certain conditions, are nearly as slippery as ice, even in the warm weather, and on which even a man, who has neither a weight to carry nor a load to draw, can hardly keep his footing. Manchester, for example, in dull, misty weather has a faculty of producing, in combination with smoke, a sort of greasy mixture, which feels to the feet as if the road and pavement had been coated with lard. This greasiness is fearfully trying to horses, especially to the magnificent animals which draw the great 'lurries.' In spite of the deep calks and toe pieces with which their shoes are armed, and which make them look as if they were walking on pattens, they slip and slide about in a most pitiful manner, the terror which they are suffering being too plainly visible in the expression of their eyes.

Yet I have handled horses belonging to a Manchester physician, which trotted freely and easily over this slippery surface, merely because their owner was wise enough to leave the hoof as nature made it.

If any road ever made by man could damage a horse's hoof, those of Manchester would stand pre-eminent. On account of the enormous weights which are carried on the lurries—*i.e.* huge, broad-wheeled waggons, the roads have to be made of corresponding strength. Beneath the actual pavement there is a thick layer of concrete, on which are laid the granite blocks which constitute the roadway.

These blocks are shaped like gigantic bricks, and are laid on their edges, the upper angles being bevelled off, so as to leave a deep groove round each stone. These grooves are intended to prevent the horses from slipping, the calkins and toe pieces being supposed to catch in them. The effect on wheels and shoes can only be realised by actual experience.

As to wheels, any one who has been obliged to take a cab in Manchester, and to drive fast in order to catch a train, will not be very likely to forget his experience for the next day or two.

As to shoes. I have before me a lurry-horse-shoe which I procured in Manchester. It has been fitted with a bar-calkin rather more than two inches in depth, while the front of the shoe had a toe-bar of the same depth and seven-eighths of an inch in thickness. Yet, the toe-bar, after about three weeks' wear, has been so rubbed down that scarcely a quarter of an inch remains.

Still, even this exceptionally hard and rough paving did not injure the hoofs of horses which were unprotected with iron shoes.

Very smooth pavements, such as those of wood and asphalte, especially in sloping ground, are apt, particularly when wet, to cause horses to slip and fall. A very familiar example of such a pavement is to be found in Ludgate Hill, which is the scene of more accidents to horses than perhaps any other track of similar length in the whole of London.

Ludgate Hill is the terror of drivers, who speak of it as if it were Mont Blanc. Yet, it is not at all a long nor a steep hill. The length is only about two hundred yards, and the rise not quite nine yards—a mere trifle. Yet, in damp, and especially in 'greasy' weather as it is called, horses are constantly falling upon Ludgate Hill.

Fallen horses are so common that few people take the trouble to stop and look at their hoofs. If they were to do so, the cause of the fall would be at once evident.

The natural safety-pad of the hoof, which would

have clung to the smooth and slippery surface, has been cut away by the farrier, while the under surface of the shoe has been rubbed down by friction, until it offers only a flat polished metal surface, seldom less than half an inch in width, and often more. Of course the shoe slips and the horse falls. The india-rubber pad would have given the horse a firm footing, and much more so would the natural safety pad or frog.

A similar sight to that which is described below may be seen on Ludgate Hill on any wet day. The writer is Lieut. W. Douglas, late of the 10th Hussars, and therefore a competent observer. The passage occurs in the preface to his book on horseshoeing, and is quoted by ' Free Lance ' in 'Horses and Roads : '

' Passing down Ludgate Hill one day, my attention was drawn to the pitiful condition of a horse in the shafts of a large waggon. The poor animal was not drawing the load, but was being driven down the descent by the crushing weight behind ; and utterly unable, from the manner in which it was shod, to withstand the pressure, it had gathered its hind legs well under and its fore legs in advance of its body, in a hopeless struggle to avert the fall which it too evidently knew was at hand.

' Never did I witness such a picture of powerless terror as that horse presented, as with eyes starting, body shaking, and knees stiffened, it was carried downwards against its will, until, the fore and hind feet slipping in the same direction, it came down upon its left side with a crash. The thought of what that poor beast must have suffered, even before it fell, has haunted me ever since.'

The waggon in question seems to have been unprovided with a brake, but Mr. Douglas is of opinion that if the horse had been allowed to possess its frogs in their natural state, it could even have controlled the pressure from behind.

I never realised the value of the frog on a smooth, wet, sloping wooden surface until my second voyage across the Atlantic.

On the outward voyage I had been greatly inconvenienced, not to say endangered, by the slipperiness of the deck and the soles of my boots. 'Sea-legs' are very useful in their way, but when the vessel rolls, sea-legs are quite useless in preventing the voyager from sliding down the deck if he be above, or down the saloon floor if he be below, and in all probability damaging himself seriously by being flung against the gunwale or the saloon furniture.

So, just before the return voyage, I had a slight layer of vulcanised indiarubber, not thicker than an ordinary playing card, affixed to the soles of the boots, shoes, and slippers which I meant to wear on board. The effect was almost magical. We had a very rough passage, but even when a promenade on deck looked like walking on the wet slates of a house-roof, the indiarubber clung to the smooth boards of the deck above, or to the carpet of the saloon below, and I never once had a fall.

This experience enabled me to understand practically what is the worth of the frog to the horse, and how the working power of the animal is diminished by removing the frog, and throwing the weight of the horse upon the flat, smooth surface of the iron hoof.

As to the calkins which are so persistently employed as preventives of slipping, I shall soon have something more to say of them.

It may seem almost incredible that the hoof of the horse can be strong enough to resist the hardest or the stoniest road, and yet be so wonderfully constituted as to be an organ of touch. Yet, as will be seen from the following narrative, such is really the case.

Some five years ago I had the pleasure of making the acquaintance of Mr. John Bellows, of Gloucester, the author of the celebrated French dictionary. He is a member of the Society of Friends, and one of the few who retain the phraseology of George Fox's day. Like most members of the society, he has a great tenderness for animals and especially for horses, and has taken a pleasure in treasuring authentic anecdotes of his favourites. He told me the following narrative, and at my request committed it to writing.

The reader will pardon the insertion of the last clause, but it is too characteristic to be omitted :—

East Gate, Gloucester : December 4, 1879.

'When my father was quite a little child, of perhaps three or four years old, he horrified his mother by trotting across the street in front of their house, in the village of Bere Regis, and tripping up exactly in front of a team of horses drawing a heavy waggon.

'As he fell, the leader horse set his great hoof on the child's head, and his mother expected to pick him up dead. But no! He used to say that all his life afterwards he kept a most distinct remembrance of the soft and gentle touch of the horse's foot pressing him to the ground "like a sponge," and holding him there until his mother relieved him.

'The horse had pulled up in the twinkling of an eye, and brought the rest of the team to a standstill. But he knew that the child was safer lying still than wandering among heels and wheels; and there he kept him, or *I* should not at this moment be able to sign myself

   'Thy sincere friend, and father of
    several of thy admirers,
      ' JOHN BELLOWS.'

In this case, the proceedings of the horse were the more remarkable, because the hoof was shod, and therefore its sensitive capacity must have been cramped. I am always glad to receive information from members of the Society of Friends, because their scrupulous regard for truth and avoidance of exaggeration gives to their narratives an additional value.

'Free Lance' puts the point very well in the following words: 'The frog is a natural calk, but it must have fair play. It is pointed in front like a ploughshare to offer resistance in one direction. To offer resistance in the contrary direction it is semi-cloven, and thus it offers a double resistance, for the very evident reason that a horse needs more aid to go ahead than he does to stop himself. Yet the two ends have been rightly balanced by nature, if we could only see the thing as such.'

So much, then, for the frog and its value to horse and therefore to man. The idea of paring it for the purpose of stimulating its growth is ludicrously absurd. The best way of stimulating the growth of the frog or of any other organ is to let it do its proper work without hindrance.

There is another objection to cutting away the frog, but this, as it relates to internal structure, the illiterate farrier is not likely to know, nor to understand it if he be told.

The navicular bone, the importance of which in the mechanism of the foot has already been described, lies between the arms of the V-like frog, and if the frog be weakened by the knife, a wrong strain is thrown on the navicular bone, which is then tolerably sure to become the seat of inflammation. Navicular disease is never found in the wild horse, nor in our own horses until the farrier has worked his will with their hoofs.

## CHAPTER VI.

Horseshoeing on 'improved principles'—Hot-fitting—'Clips' and their origin—Groove cutting in the hoof—Natural thatch of the coronet—Natural varnish of the hoof—Abuse of the rasp—Blacking hoofs—Effect of grease on the hoof—'Stopping' hoofs—The 'bottle of oils' —Its effect on the hoof—Drugs in stables—Horses poisoned—Thirty thus lost by one owner—Anti-drug Association—The rashness of ignorance.

On account of the extreme importance of the frog, it has been necessary to give a considerable amount of space to it. We will now revert to the 'improved principles' of horseshoeing.

After having seen the havoc which has already been made in the hoof by the farrier, it might reasonably be doubted whether more mischief could possibly be done, and whether the hoof could be further injured. Human ingenuity, however, proved itself equal even to this task, and 'hot shoeing' and 'clipping' were invented, both, as usual, for the benefit of the horse.

'Hot-fitting' or shoeing is thus accomplished.

'In fitting the shoe, the coaptation between it and the hoof should be as close as possible. This can most readily be secured by applying the shoe at a high temperature, and for the briefest space of time, to the part on which it is to rest. By this means the inequalities in the horn can be perceived, and removed by the rasp, and when quite level, another brief application of the hot shoe fuses the horn into a hard level surface, capable of resisting the pressure of the metal during wear.

'This " hot-fitting" of the shoe, as it is termed, is perfectly harmless to the *unmutilated* hoof, and possesses such great advantages that it is to be commended. By " cold-fitting " it is impossible to obtain such an intimate coaptation ; and even if it could be secured, the shoe would not remain so firmly attached, as wet softens the ends of the horn-fibres in contact with the shoe, and they yield to the pressure, the shoe loses its original bed, becomes loose, and is cast. This is the experience of those who have tried this kind of fitting most extensively.'

These directions were written by one of the first authorities in veterinary science—namely, Professor G. Fleming, veterinary surgeon to the 2nd Life Guards. They occur in an article written for the 'Live Stock Journal Almanack' of 1879.

Any opinion of such a writer is worthy of respect, but when I read these words for the first time, I could hardly believe my eyes. Of course, if time were an imperative object, hot-fitting would be a quicker process; but where a few minutes are of no particular importance, it must be injurious to the hoof, without ensuring any corresponding advantage.

I did not like the repeated use of the word 'co-aptation,' just as I always suspect the practical knowledge of a medical man when he speaks of cephal-algia being a premonitory symptom of incipient rubeola, when he might just as well have said that headache was one of the signs of measles.

Another point struck me—namely, that the 'bed' obtained by the red-hot shoe would *not* fit the same shoe when it was cold, owing to the great expansiveness of iron when heated. Again, though the application of the red-hot shoe would not give pain to the horse at the time, any more than we should suffer pain if the tip of an overgrown finger nail were scorched, yet the horn of the nail would be rendered brittle for some distance beyond the portion that was actually burned away, and the same would be the case with the hoof of the horse.

Lastly, the reader must have noticed that the writer seems to have had his misgivings about the universal employment of hot-fitting, and carefully limits it to the *unmutilated* hoof. The italics are his own.

I believe that Mr. Fleming has since abandoned the practice of hot-fitting, but I have quoted the passage in order to show how curiously habit will overpower reason, even in a well-trained mind.

His saving clause of the *unmutilated* hoof is quite sufficient to show that the practice is indefensible. Where do we ever meet with an unmutilated hoof except in the case of young horses who are to be shod for the first time. Even in their cases the hoof is not in its normal condition, for the animal has passed all its life on the comparatively soft and smooth surface of a paddock, and the hoof has not been sufficiently hardened to endure its proper work.

HAVING now scooped the sole to the thinness of paper, cut away the frog, removed the pegs, lowered and cut open the heels, and burned away the wall with red-hot iron, so as to make it brittle, and all for the benefit of the horse, the farrier has yet another resource for weakening the hoof with his ever-ready knife.

In order to save the insertion of more nails than necessary into the horn, which has become honey-combed by nail holes, 'clips' have been invented.

These are simply flat pointed projections from the shoe. They can be hammered over the hoof while cold, and serve to hold the shoe to the hoof without the use of nails. Clips are no new invention, and indeed were used for fastening the shoe to the hoof long before any one dreamed of hammering nails into the horn. In fact, the original iron shoe was simply a thin flat ring of iron with three clips, one coming over the toe, and the other two over the quarters.

If, then, the simple clip were used, no direct harm would be done to the hoof. But the farrier

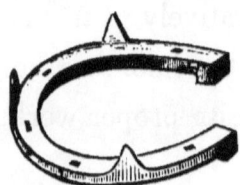

SHOE WITH CLIPS.

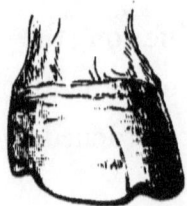 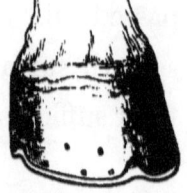

HOOF SCOOPED FOR CLIPS, AND CLIPPED SHOE ON HOOF.

likes to see a 'clean' hoof, and thinks that the look of it would be injured because the clips would slightly project, so he scoops a semicircular groove for the reception of the clip.

These shoes which are furnished with clips always have one in front, so that scooping the groove entirely destroys the original contour of the hoof. Moreover, the hoof is weakened exactly where it wants most strength, and what with lowering the wall until it is perilously near the linea alba, scorching it brittle with red-hot iron, and then scooping a groove in it, the farrier has done his best to produce seedy toe, gravelling, and consequent founder.

As to cutting a groove to receive the clip on the quarters, it may be urged that if the inside clip were not sunk into the hoof it would cut the opposite leg. So it might; but that is no reason for sinking either the toe clip or the outside clip into the hoof, even granting that either were necessary. But in the eyes of the ordinary farrier external show and the fashion of the day are everything, while knowledge of anatomy, reason, and common sense count for nothing.

So wedded are the farriers to custom, that instead of using the clips as a succedaneum for nails, they use the clips and the nails besides. There is now before me a horseshoe which I purchased from a well-known forge, and which not only has three large clips, but *ten* nail-holes besides, so that the grooves for the reception of the clips not only weaken the wall, but diminish the space in which nails can be safely inserted.

As to the nails themselves, they will be presently mentioned when the shoe comes under consideration.

Two more methods of injuring the hoof are at the farrier's disposal after the knife and the red-hot iron have been employed. He has yet at his command the scissors and the rasp.

The reader will remember that the wall of the hoof is secreted from the coronary ring, or band, which is necessarily a vascular and consequently a sensitive structure, and the blood, by passing through it, is converted, by a chemical process far beyond the reach of man, into horny fibres. It is, in fact, analogous to the root of the human nail. In order that this structure shall be protected from rain, snow, and wind, it is furnished with a sort of a roof or penthouse of hair. For some reason or other, the farrier dislikes this covering, and therefore cuts it away with his scissors.

Then the rasp has its turn. First, the man goes round the edge of the hoof, so as to file the hoof level with the shoe. It is so easy to do this, that one writer on horseshoeing, being aware that the farrier *will* rasp the edge out of mere habit, recommends that the shoe should be made a little larger than is needful, so that the shoe may be filed level with the hoof, and not *vice versâ*, as is mostly the case.

Having filed all round the hoof, the farrier next thinks it necessary to go over the entire surface.

Nature has covered the hoof with a sort of shining varnish, which permits the passage of air, but yet defends the structure from wet. This varnish seriously offends the eye of the farrier, who proceeds to remove it with his file. I have before me a number of hoofs, and only one of them does not show file marks over the entire surface. That hoof, however, belonged to an unshod Circassian horse, and man has never tampered with it.

THE farrier is at last satisfied with his work. He has made 'a nice, clean foot,' and the animal is taken away to its stable. Now the groom has *his* turn at spoiling the hoof, and all with the best intentions towards the horse.

Hoofs are naturally mottled with various shades of brown, grey, and black, while in some places they may be nearly white. This mottling, however, does not please the eye of the groom, who considers that the hoof of the horse and his own boots ought to match each other. So, with grease and lampblack, he blackens and polishes the hoof, and then thinks that he has done his duty by his horses. It is a very dirty and disagreeable process, so the very fact that he does undertake it shows that he really thinks that the horse is improved by it.

True, for a short time, and if the road should happen to be in good condition, the hoofs retain their polished blackness; but if the road should be dry, they will soon be covered with dust, which sticks to the greasy surface, and gives it exactly the neglected and untidy appearance which the groom has tried to avoid.

Then, after the horse has been taken back to the stable, cleaning the hoof is as disagreeable and dirty a process as blacking it, whereas dust cannot adhere to the natural varnish, and the hoof can be easily cleaned by a wet cloth.

This, however, is not the worst result of grease and lampblack, which would work no great harm if they merely caused additional trouble to the groom, but absolute damage to the hoof is done by the grease. The groom, placing the hoof of the horse on a level with his own boots, has no idea that the former needs more consideration of treatment than the latter. He has no idea that transpiration takes place through the hoof, and that the grease chokes up the pores in the horn, and so prevents the needful transpiration from taking place.

As might be expected, the stoppage of these pores, and the check to transpiration, causes the hoof to become hot and feverish, just as is the case with ourselves when perspiration is checked. The horn, too, cannot be properly formed, and becomes brittle.

Then the groom sets to work to 'soften' the hoof, which is the very thing that ought not to be done, and proceeds to 'stop' it—*i.e.* to fill the cavity with some mixture which has been recommended to him, and on the virtues of which he implicitly relies. Fresh cowdung is the usual material employed for stopping hoofs, the groom having an idea that because it is soft it is cooling, whereas it really is heating in its effects.

Then, most grooms possess some special hoof-ointment, and there is hardly one who has not a 'bottle of oils' to which he pins his faith with a blind trust deserving a better cause. How the 'oils' are supposed to act on the hoof he does not know, nor care to know, and, as a rule, he is personally affronted if any one wishes to teach him anything of which he is ignorant.

So he goes on stopping the hoofs with his mixtures, and dressing them with his ointments and oils, which, as Mr. G. Ransom says, 'are used more openly than medicines, and are even highly approved by some owners.

'First among them rank hoof-ointments, be they either a secret with the stablemen or a patent. It does not make much difference which, as to their non-ability or rather their positive insalubrity. They almost always consist of admixtures of some or all of the following ingredients: Tar, bees-wax, train-oil, tallow, suet, and honey.'

He then proceeds to endorse a statement made by another authority, who boldly says that if these applications were made daily and not occasionally, the horse would not in six months' time have enough sound horn left in the hoof to hold a nail.

The injury which is done to horses by the well-meant but reckless treatment of their hoofs by ignorant and therefore conceited men is so great, that no groom or stableman ought to be allowed even to possess a hoof-ointment, or 'oils,' or any similar preparation, or to apply anything to the hoof which he has not received from the veterinary surgeon whom his master employs, and in the application of which he does not strictly follow the surgeon's orders.

It is necessary that the rule against the possession of drugs or applications of any nature should be enforced most rigidly, and that the slightest infringement of this rule should be visited by instant dismissal and forfeiture of wages, without the hope of forgiveness or of being reinstated at any future time. No other plan is of the slightest avail, and every servant who is employed about the horse in any capacity should be compelled to sign a clause to that effect before he is engaged.

Not only do the men apply remedies to the hoof, but they procure and administer drugs of the most poisonous character. Their rashness and presumption are almost incredible. Here are two extracts which I cut from the police columns of a daily paper:—

'*Caution to Farm Servants.*—On Monday last a waggoner in the employ of Mr. H—— was charged before the magistrates with administering poisonous drugs to his master's horses, and was sent to gaol for one month with hard labour. The poisons consisted of savine and sheep salve. We hope this notice will act as a warning to others holding similar situations, and who, we fear, are too often guilty of the like dangerous practice, for the sake of making the horses look bright, at the expense and risk of their valuable lives.'

'At the Petty Sessions to-day, W—— J——, servant to Mr. H——, farmer, was charged with administering a poisonous compound to his master's horses, and on conviction the magistrates fined him 3*l.* An expert stated that antimony was the principal ingredient, and that the poison had been given to produce a glossy skin. Three of Mr. H——'s horses had died through this practice.'

The reader has doubtless noticed that in neither of these cases was the horse intentionally injured. Had it been so, the punishment would have been much more severe. So far from wishing to be cruel to the horses, the men evidently wished to benefit them, and only treated the animals with the same recklessness with which they would have treated themselves or their friends.

I have had large experience among the poor and uneducated, and have always been struck with the curious fact that as regards medicines they will much rather take the opinion of persons in their own rank of life than that of the President of the College of Physicians, and never seem able to overcome a lurking distrust of any one who is better educated than themselves.

Arsenic, corrosive sublimate, nitre, aloes, and drugs of a similarly powerful character are among the medicines which the groom likes to have by him, and which he will administer as often, and in such quantities, as may happen to please him. So the only plan is not only to forbid to the servants the administration of any drugs or applications, but to make even the possession of them a ground for instant dismissal.

'So nearly related,' writes Mr. J. Irvine Lupton, 'are the quantity of aloes which relaxes and the amount which kills, that probably aloes have poisoned more horses than all other drugs in the Pharmacopœia.' After entering into the anatomical details of the horse's digestive organs, and the effect of this drug upon them, he proceeds as follows:—

'Aloes and nitre are the chief perils of the stable.

'More horses have diseases of the kidneys through the abuse of nitre than would be effected if left entirely to Nature. As to aloes, the poisonous and perilous nature of that drug has already been dwelt upon; the pitiable infatuation with which grooms regard it constitutes one of the heaviest and needless extravagances of every hunting establishment.'

Mr. Ransom mentions that in 1879 a number of Lincolnshire farmers met together and formed themselves into a society for suppressing the administration of poisonous drugs to horses by servants. One of the members stated that he had lost more than thirty horses by this practice.

It would have been better if the word 'poisonous' had been omitted, and that all drugs had been prohibited.

Ignorant men cannot be expected to discriminate between poisonous and harmless drugs, nor to be at all nice as to the amount which they administer or the ailment for which they administer it. To them a drug is 'physic,' and therefore a panacea, and they will with astonishing complacency administer the same 'physic' to a horse, or a cow, or a pig, or themselves, or their babies, without the least reference to the composition of the 'physic' or the nature of the ailment.

A mother, for example, will find that her baby is not looking well, and thinks that it wants 'physic.' So she remembers that last year her husband had some physic from the dispensary, but did not quite finish it. Accordingly, she had put the remainder away, lest it should be wasted, and now gives it to her baby, who may be suffering from teething, or whooping cough, or measles, or chicken-pox. Any experienced dispensary surgeon has seen plenty of such cases.

The clergyman of the parish would not dare to do such a thing, because he is sufficiently educated to know and acknowledge his own ignorance of medicine. It must be evident, therefore, that half-measures are useless, and that if a horse-owner does not wish to have his horses poisoned by drugs or lamed by ointment and oils, his only plan is to forbid them absolutely.

## CHAPTER VII.

The shoe—Artificial roads and artificial protection—Variety in roads—Straw shoes of Japan—Raw hide or 'parflèche' shoes of North American Indians—Shoe nails—Their ordinary size and number employed—Diminishing thickness of wall—An old Scotch law—The 'unilateral' system—A hoof prepared on the 'improved system'—A mangled hoof restored by Nature—The dangers of shoe nails—Cut nails and forged nails—A remarkable accident—Effect of a heavy shoe on the horse—'Marden' and the dead heat—Effect of a heavy shoe on the muscles—Lancashire clogs and French 'sabots'—Cetewayo and followers in England—The 'lurry' horses and their shoes—Lieut. Douglas's calculations—Loosened hoofs.

HAVING now treated of the hoof, we will pass to the shoe.

The object of the shoe is to benefit the horse by protecting its hoofs against the hard artificial roads of civilised countries. It is always assumed that such protection is necessary, because, although the horse might do very well without these appendages in its native pastures, it must need shoes when it is brought under such different conditions, and has to traverse stone-paved roads when it carries a rider or draws a vehicle.

If such assumption be justified, all we should

have to do would be to invent a shoe which will protect the whole of the hoof and at the same time will not interfere with its proper functions. But, at the very outset, we are met with difficulties. If all roads were alike, nothing could be simpler, and, as one writer observes, 'The ingenuity of man would devise horseshoes to travel over glass, were glass the only pavement in use.'

But all roads are not alike. There are hilly and level roads, and even these are not alike. Nothing can be more dissimilar than the chalk and flint-paved hills of Derbyshire, or the hard stony hills of North Staffordshire. The level roads of flat country have all their distinctive points of dissimilarity, and so have the roads of cities; the hard, uneven granite-paved roads of Manchester, for example, having little in common with the asphalte and wood of the London streets. Again, all hoofs are not the same in quality, some being hard and tough, while others are weak and brittle.

Possibly for these reasons, the variety of horseshoes that have been produced by 'the ingenuity of man' is beyond all calculation. One man alone has invented twenty different forms, a tolerably good proof that nineteen of them were faulty, and would damage rather than benefit the horse.

A very important point in horseshoeing is the

method by which the shoe is attached to the hoof. It must be clear to every one that in proportion as the hoof is injured by the operation, the shoe must be a bad one. The Japanese have a very simple and ingenious mode of shoeing horses. The animals are not shod at all, according to our ideas—that is, the shoe is not attached permanently to the hoof.

But, on a journey, a rider or driver is provided with a number of shoes made, not of iron, but of straw. They are not made only to surround the hoof, so as to throw the weight of the animal on the wall, but are, in fact, thick sandals made of closely plaited straw, and tied on the foot by thongs.

Their cost is a mere trifle, and when they are worn out they are thrown away and another set substituted for them if needed. They are only used when required, so that the sides of a bad piece of road are always strewn thickly with worn-out sandals, while scarcely any are to be found on the good portions.

In North America, shoes of a similar nature are employed. They are not, however, made of straw, but of the all-useful raw bison hide called 'parflèche,' and, except in shape, are identical with the moccasins which are used by the hunters, whether they be white or red men.

As to iron shoes, the only kind which does not

injure the hoof is the three-clamped ring which has already been mentioned. Nails, however few or slight, however well they may be made, or how skilfully they may be driven, *must* injure the wall, even if it be perfectly sound and hitherto untouched.

If the hoof were a mere solid, self-renewing block of horn, without any particular structure, no very great harm would ensue, as the nails would only make holes which would be soon filled up again. But the only part of the hoof into which a nail can be driven is the wall, which is made, as has been said before, of flat fibres laid side by side.

Now, all shoenails are made on the principle of the wedge, and if a wedge be driven between parallel fibres, it must tear them apart. The rent thus made is of course most conspicuous at the nail-hole, but is necessarily continued above and below it.

On an average, eight nails are used in a shoe, and on an average each nail is at least one-eighth of an inch in thickness. So, by inserting these nails, the farrier is driving a full inch of iron among the fibres of the wall, tearing them apart at the nail-holes, and crushing them together at each side, thus destroying the whole character of the horn. Then, after the shoes have been removed and replaced a few times, the horn becomes so weakened that it will not hold nails unless they are driven higher into the wall, where sound horn is still to be obtained.

Again, this practice would not be very hurtful were the hoof solid, or if the wall were the same thickness throughout; but, as the reader may see from the section figured on p. 40, the wall diminishes in thickness up to the coronary ring, where it becomes a mere edge of horn.

Not only does it diminish in thickness, but in strength, so that just below the coronary ring it is as comparatively soft as is the horn at the base of our nails. It also increases in sensitiveness, so that there is very great danger of 'pricking' the vascular portions, and therefore of laming the horse at the least, even if worse results do not ensue. Many cases have been known where the result of a prick has been the death of the animal through lockjaw.

In former times there used to be a law in Scotland that if the shoer pricked a horse, he was obliged to nurse the animal until it had recovered, and to provide an efficient substitute as long as the victim of his carelessness was unable to work. If the horse died from the wound or its results, an equally good one must be given to the owner. I very much wish that such a law could be re-introduced and acted upon throughout the United Kingdom.

There is now before me a hoof, kindly presented to me by Mr. G. Ransom ('Free Lance'), in which one of the nail-holes is exactly an inch and a quarter above the edge of the hoof, so that the nail must have passed quite close enough to the vascular laminæ to cause pressure on them, and therefore to give pain.

This hoof is peculiarly instructive, because it has been shod on what in veterinary language is called the 'unilateral system,' in English the one-sided plan, the shoe having been nailed only on one side and on the toe. I have exhibited it throughout England, Scotland, and a great part of America, and have always found that the spectators were greatly struck with the torn and battered horn of the nailed side, and the clear, firm, and beautifully grained horn of the untouched side.

Originally, the shoes had been nailed all round as usual, but on the sound side all the nail-holes have disappeared, with the exception of one, whose position is externally indicated by the crack in the fibres above it. The person who prepared the hoof has unfortunately polished and trimmed it, but the damage which he has thereby done to the hoof as a specimen is partially compensated by the prominence which is given to the 'grain' of the fibres, and the ease with which an injury to them can be traced.

I possess another hoof still retaining the shoe. On looking at the interior of the hoof, it may be seen that one of the nails has passed through the linea alba, or quick of the nail, so that the pain which the horse must have suffered can hardly be imagined.

I have here assumed that the farrier is putting a shoe on a hoof that has not been touched by the knife. Even in such a case, the greatest care is required in order to avoid accidents. But, suppose that even the most skilful farrier in the world were required to put a shoe on such a hoof as that which is shown in the accompanying illustration, he would find the task almost impossible.

HOOF OF MR. HERBERT SMITH'S HORSE.
(See page 188.)

Some persons might say that this drawing was heightened by the artist in order to produce more effect.

Now, there is nothing more abhorrent, even to a novice in any branch of science, than 'heightening' or 'colouring' a statement. But, in order to avoid even the suspicion of colouring, the drawing is an exact copy of a photograph which was sent to me by the late Mr. A. F. Astley, who also furnished me with the name of the owner. The hoof in question is that of the left fore-foot, and that of the right fore-foot had been quite as badly treated by the farrier's knife.

By way of contrast, I give a figure of a hoof as it ought to be. This is taken from another photograph sent to me by Mr. Astley. The hoof had originally been as badly maltreated as the former specimen, but when released from the knife, the rasp, the nail, and the shoe, it returned in a few months to its original shape.

I especially wish the reader to remark that throughout this work I place actual facts before him, and do not air my own theories, fancifully illustrated by an obliging artist.

HOOF, NEARLY PERFECT, MR. LUCK'S HORSE.

There is a peculiar danger about shoe-nails which is little suspected. 'Cut' nails, as everyone knows, are far cheaper than forged nails, and for many purposes, especially in carpentering, are quite as useful. But, for horseshoeing, they are terribly dangerous. Cut nails are made by rolling sheets of iron to the requisite thickness, and then punching the nails out of the sheets, much as steel pens are made, and as military gun-caps were made before the breech loading rifle came into use. Now, the effect of this rolling and punching is to produce 'lamination'—*i.e.* the iron forms itself into layers. These layers do not show themselves in the nail until it is used. Then, however, the blows of the hammer cause the laminæ to separate, and so it happens that the nail has a double or even treble point. In one case a nail acted in a most curious manner. It was rightly placed, and came out on the exterior of the hoof, just where it was expected. But the horse went lame after shoeing, and when the shoe was removed in order to find the cause, a most unexpected accident was discovered. The nail had split so as to look like the capital letter Y. One-half had followed the right direction, but the other had turned inwards and passed through the edge of the coffin bone a little way from the toe, so that it lodged among the sensitive laminæ.

If this had been all, the horse might have been saved, but in removing the shoe in the usual manner, the inner fork of the Y was broken off and left in the toe. The result was that in a few days the horse died from lockjaw.

No fault was to be found with anyone, as the nail was correctly driven, and to all external appearance everything was right. That the lamination produced by rolling the iron into sheets and increased by the punching of the nail should lead to such a result could not have been expected at the time, though for the future any farrier who wilfully uses cut instead of forged nails ought to be held responsible for any damage to the horse.

While I was in America I saw some horseshoe nails which were beautifully made. They were hand-forged, and so tough that they could be twisted when cold into a screw, and then twisted back again without breaking or even showing a crack. They were of course more expensive than the rolled and cut nails, but if nails similarly made had been used in the above-mentioned case, the horse would not have been lost.

These nails were shown to me in a railway car, and I should have liked to have secured a few specimens, but could not do so. But I have an indistinct idea that they are called Putnam nails—at least, there is a nail which goes by that name, and which seems, from its description and appearance, to be similar to, if not identical with, the nails which I saw.

Another evil of the shoe is its weight. What effect that weight has upon the sinews and muscles of the leg has already been mentioned, as well as the needless and additional work which is thrown on the animal.

We feel it ourselves when we run, and all who have given any time to foot-racing are practically aware of the necessity of having the shoe and running sock made as light as possible. Indeed, some pedestrians dispense with even the light shoe and sock, and run barefoot. A valuable proof of this fact occurred in 1882. In the 'Sandown Derby,' a most remarkable event occurred, no less than three horses coming in together and running a dead heat. Of course the dead heat had to be run out, and all three horses started for it. During the interval, the owner of one of them, named 'Marden,' took off even the light plates which are substituted for shoes before a race. The result was, that Marden won easily by three lengths, the removal of a few ounces of weight from its hoofs having given the animal an advantage equal to the deduction of as many pounds from the weight of the rider.

As to the injury caused to the muscles and sinews of the horse by depriving the hoof of its proper play, and forcing the animal to lift a useless weight from the ground, we may find a parallel among ourselves.

If we look at the legs of a country 'clod-hopper,' or of a Lancashire mill-hand in England, or of an ordinary labourer in France, we shall see that from the knee to the ankle there is scarcely any development of muscle, the calf of the leg being almost entirely absent. This fact has been very well hit off by Tenniel in his illustrations to 'Alice in Wonderland,' the legs of 'Father Williams'' son being curiously true to nature.

This want of development is due to the structure of the boots worn by rustics. They are exceedingly heavy to begin with, and are made heavier by the soil which clings to them. Then the soles are so unbending that the instep has no play, and therefore the muscles of the calf which work the instep have so little to do that they cease to be developed.

The wooden soles of the Lancashire clogs, and the wooden shoes or 'sabots' of the French rustic, produce a similar effect, while exactly an opposite result is found in the professional dancer, the muscles of the calf being enormously developed. The 'light fantastic toe' is unknown to the field labourer, whose only idea of dancing is a shuffling clatter; and for the same reason, leaping is to him almost impossible.

When Cetewayo was in England, he and his followers found the weight of their shoes so fatiguing, especially when ascending stairs, that after they had been for a visit to any place where stairs had to be ascended, they were so worn out that they were obliged to pass a day or two prostrate on their strawbeds.

As with us, so with the horse; and every ounce

which is hung upon his feet adds greatly to his day's work, and helps to wear him out before his time. Common sense ought to tell us that, even regarding the horse as a mere machine, it must be as foolish to hang needless weights on his feet as to clog an engine after a similar fashion.

Now, as a rule, the weight of a shoe for a carriage-horse is about a pound and a half. Multiply this weight by four, and we can at once see how much needless work has to be done by the mere exertion of lifting such a weight from the ground.

As extreme cases, I may mention the shoes which are worn by the gigantic horses employed to draw the Manchester 'lurry.' An ordinary set of 'lurry' horseshoes weighs about sixteen pounds, this enormous mass of iron being thought necessary to protect the hoof against the granite pavement.

There are some sets which weigh seven pounds per shoe—*i.e.* twenty-eight pounds per set. These, however, I never saw, but accept the statement of Lieutenant Douglas, in his 'Horseshoeing,' p. 4.

The same writer makes a curious calculation of the difference in a horse's work when he is shod lightly or heavily. The directors of one of our large railway companies substituted shoes weighing ten pounds per set for those which were double the weight.

'And what was the result? Why, that the light shoes wore longer than the heavy ones had done, the average of the former being four weeks and five days each shoe, while the latter wore but three weeks on an average.

'But it is not the *wear of iron* so much as the *wear of horseflesh* that I am anxious to direct attention to now. It will be seen that as the shoes which were taken off weighed five pounds apiece when new, these horses were at once benefited by having ten pounds less iron to carry about with them.

'I will not attempt to calculate what difference two and a half pounds of metal placed at the end of a lever two and a half feet long would be equivalent to, but will simply look at it as ten pounds extra weight carried by each animal. A horse when walking lifts his feet all round about thirty times a minute, so that if we allow the day's work to last eight hours, the difference in favour of lighter shod horses is over sixty-four tons.'

Here Mr. Douglas gives his calculations in full; but, as space is valuable, I do not transcribe them.

The heavy unyielding shoe has sometimes a very strange effect. I have seen a 'lurry' horse, who had not acquired the art of lifting his weighted feet perpendicularly, try to walk as it would have done had he been unshod—*i.e.* by putting down the heel first, and ending with the toe. The result was that the hoof was becoming loosened from the foot; and when the horse raised its foot, the bones and their sensitive laminæ were drawn out of the hoofs so far that the vascular portions became visible, the foot sinking back into the hoof when the weight of the horse rested upon it. The agony which each step must have cost the horse is beyond the power of description.

## CHAPTER VIII.

The calk, or calkin—Horses on pattens—Two strange accidents—Calks in America—Supposed uses of the calk—Mr. Bowditch's testimony—Weight thrown on the edge of the coffin bone—High-heeled boots and their effects—The battle of the shoes—Recognition of defects in shoeing—The Goodenough shoe and its object—Jointed shoes—The Clark jointed shoe—The screw shoe—Expansion and contraction—The effect of the screw on the hoof—Tips, and how to fasten them—The Charlier, or 'pre-plantar' shoe—How to apply it—Usually too large and in danger of breaking and twisting—Best length and weight for a Charlier shoe—Man *versus* Nature—A series of happy thoughts—Their results upon each portion of the hoof.

THERE is one portion of the shoe which must be mentioned. It is the calk, or calkin—*i.e.* a projection at the heel which looks very much like the high-heeled boots which have destroyed the feet and broken the health of many a fashionable beauty. Sometimes, as in the Manchester shoe, which I mentioned on p. 113, the calkin extends across the back of the shoe, connecting the ends together, and looking very much as if the blade of an iron scraper had been fastened across the heels. Generally, however, there are two calks, one on each side of the heel; and, too often, the blacksmith has not troubled himself to make them exactly of the same height.

I never shall forget the first sight of these astounding shoes. To be able to see *under* the feet of a horse was an unknown experience, and in this instance the effect was heightened by the rays of the setting sun, which produced the most curious combinations of light and shade as the horses raised or set down their feet.

Lieut. Douglas, in his usual incisive style, is very emphatic in the condemnation of these shoes and of calks altogether. After describing the sufferings of London horses whenever the road is hilly and happens to be slippery from any cause, he proceeds as follows:—

'It is even worse in Lancashire and other parts of the North of England, where horses are propped up, as it were, on short stilts, having calks both at toes and heels.

'Without taking into consideration the extra weight which the horse has to carry, everyone can imagine how insecure the animal must feel when walking in these tripods. We can easily fancy how we ourselves should feel if compelled always to walk about in pattens; and yet I doubt whether we should feel more miserable than these horses do.

'Some may think that I express my feelings too

strongly on this subject, but it is ever before my eyes. I cannot move out of doors without being obliged to witness horses on all sides of me suffering from bearing-reins or bad shoeing. The very day I penned these lines, when going down Cannon Street, I saw a horse fall so suddenly that the pole of the vehicle in rear passed through the back of the four-wheeler he was drawing. The poor animal never tripped; his legs flew from under him to the right, and he fell upon his left side, the wheels of the cab being only stopped by his body. The horse was shod with shoes that had high calks.

'This is no solitary instance. On the very Friday previous, when walking from Holborn Circus to Newman Street, Oxford Street, about a mile, I saw no less than five cab-horses down, all of them falling on their sides as if they had been shot dead. The streets, after being watered, or a slight shower having fallen, are as slippery as if they were covered with soft soap, and horses with high-heeled calks and wide-webbed shoes are quite helpless upon the granite pavement. These which I saw fall could not get up until rugs had been spread in front of them, so that they could get a foothold and keep it.'

Any Londoner who uses his eyes must have seen many such accidents. Two which I witnessed impressed themselves very strongly on my mind.

One was the fall of an enormous cart-horse at the foot of Ludgate Hill. The animal had struggled, and slipped, and slidden all the way down the hill, until the level ground was reached at Farringdon Street.

Just when everything was apparently safe, away flew all the four legs to the right, and down came the animal on its side with a tremendous thud, falling upon the chain trace. Instantly the animal plunged and struggled to its feet, and again fell upon the chain in a similar manner. The movements were so quick and unexpected, that the horse fell four times before anyone could approach, and so heavy were the falls upon the chain, that I doubt whether the animal survived the injuries.

The other accident occurred to the horse of a hansom cab, nearly opposite the General Post Office. It was in the winter time, and a slight thaw had been followed by a sharp frost, so that the road was exceedingly slippery. On account of the state of the road, the driver was only walking his horse, the high calks with which the shoes were supplied being of little if any use upon the icy surface.

Suddenly the horse fell as if shot, and, slow as was the pace at which the animal had been going, the fallen horse slid for some few yards, drawing the cab after it.

The most extraordinary portion of the incident was the effect upon the driver. The fall of the horse did not appear to disconcert him, and he kept to his seat as long as the cab was in motion. The moment it stopped he rose up slowly, stooped forward, and put the top of his head on the roof of the cab. Then he turned a somersault in the air, and came flat on his back in the road by the side of his horse. The deliberation of the movement was one of the most extraordinary spectacles that I ever witnessed.

I thought that the man must have been killed on the spot, or at lest stunned and insensible. But he was hardly down before he was up again. What had happened, or where he was, he evidently did not know, but he walked round and round the cab and fallen horse until the inevitable crowd surrounded the scene of accident, and shut him from sight. However, after five or six minutes had elapsed I saw him again in his seat, and driving slowly away from the stand.

The uselessness of the calking could not have been better demonstrated. It did not prevent the heavily-burdened cart-horse from falling on a wet road, nor the lightly burdened cab-horse from a like misfortune when the surface was frozen.

The object of the calking is to prevent the horse from slipping on frozen ground or ice. The same lecturer who averred that in America the farriers did not cut away the frog nor scoop the sole, also asserted that in America calks are not used.

Now I took especial care to see whether the horses wore calks or not, and I unhesitatingly say that the calks are quite as universal in America as in England, and quite as deep—possibly deeper. I asked why they were worn, and was given the usual answer—*i.e.* that without their use the horse would slip and probably fall on the sloping and frozen roads of Boston; and that horses who had no such protection did fall, and often had to be killed. Very much the same reasons are given in England.

Now, in the first place, the existence of the calk, though it may be only a short one, is tantamount to the abolition of the frog, which may just as well be cut away as left suspended above the ground. We have already seen that pressure on the frog is necessary for the well-being of the horse, so that this fact alone is decisive against the calkin.

But the assertion that without the calk the horse would slip and fall is a pure assumption. If the ordinary shoe be used, there is some reason for the statement. In winter time, as I know from personal experience, it is difficult even for a man who has nothing to draw or carry to keep his footing in the streets of Boston, and horses shod in the ordinary way would have no chance of doing their work. Calks certainly do assist the animal to a certain degree at the time, but they also inflict injuries from which the horse is seldom free afterwards, as we shall presently see. But even calks are not infallible. I once saw in New York, within a distance of two hundred yards, two horses lying dead in the road, and another so much injured that it had to be killed. In two of these cases the horse had slipped, fallen, and the calk had become jammed in the tram-rails. The result of the fall was that the pastern was snapped, and in one case the hoof was so twisted that the toe pointed to the rear.

Mr. Bowditch, an American gentleman whose acquaintance I had the pleasure of making, is one of those men who think for themselves, and have the courage to act upon their opinions, without any reference to precedent. Instead of making the shoe with calks on the heels he only fastened on the toe a slight semicircular piece of iron, leaving the rest of the hoof to grow as Nature made it. During the winter, when the roads were covered with glare ice, all the precautions which he took against slipping consisted of one small point *on the toe.*

'I am afraid,' he writes, in a letter quoted by 'Free Lance,' 'that I drive very hard down hill. I am in the habit of driving cripples; my friends have a good deal to say about the corpses that I drive, but I take care of their feet, and they manage to do good work.

'I make my best time in driving down hill. I have no fear of hard roads and no fear of pavements if a horse's foot is kept in proper condition. Last winter I rode my saddle mare, and of course my neck is more to me than anything else. I galloped out on the ice where the men were cutting it, and I had no fear of her slipping, although the horse that was marking the ice, and had calks on two inches thick, did slip.'

Finding it impossible to induce the ordinary farriers to make or put on any shoes except those of the old pattern, Mr. Bowditch boldly set up his own forge for the benefit of his own horses. Sometimes, when one of his neighbours has a very lame horse, he brings the animal to Mr. Bowditch, and when it is cured, he goes back to his old farriers, and has it lamed afresh.

Now we can see why the calks inflict an injury on the horse.

In the first place, they give a wrong bearing to the hoof, by lifting it up behind, whereas Nature intended it to come flatly on the ground. If the reader will again look at the section of the pastern on p. 40, he will see that the tip of the coffin bone resembles a wedge. Note, when the heel is elevated by the calks, the coffin bone is pressed down into the hoof, just as the human toes are squeezed into the tip of the boot when high heels are employed.

Human toes suffer considerably when thus treated, but the horse must suffer much more, because the vascular laminæ are also sensitive, and are crushed together by the weight of the animal. Moreover,

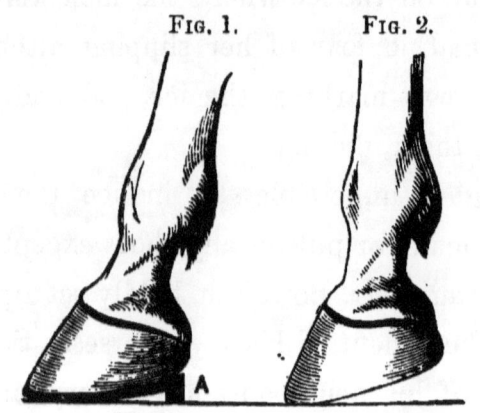

FIG. 1.   FIG. 2.

HOOF WITH CALKED SHOE.   ATTITUDE OF HOOF, THE SHOE BEING REMOVED.

however greatly human beings may suffer when wearing tight, high-heeled boots, they can at night take off the instruments of torture, whereas the horse enjoys no such respite from pain.

Another drawback to calks is the loss of power which they entail. When the hoof is tilted by the calks, the weight of the body is thrown forwards, and the muscles and sinews are strained in order to preserve the balance. This is evident by reference to fig. 2 in the accompanying illustration, while fig. 1 shows how the eye is deluded by the shoe into an idea that the hoof is on its proper level.

It is so with ourselves.

In the accompanying illustration, fig. A represents the human foot resting on the ground in the attitude designed by Nature. Fig. B shows a boot that was

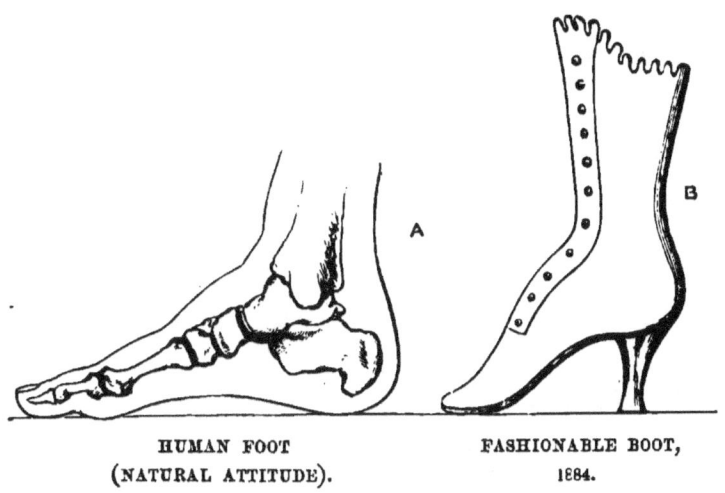

HUMAN FOOT (NATURAL ATTITUDE).  FASHIONABLE BOOT, 1884.

fashionable in 1884, and I fear, on account of its utter and fatuous disregard of every function of the foot, may retain its hold on fashion for some years to come. A mere glance at the two figures will show the loss of power consequent on wearing such a boot. The wearer is nearly as helpless as a small-footed Chinese woman. She—for I regret to say that only lovely woman can be such a slave to fashion—cannot even stand upright, still less walk upright, nor can she even stand still.

The body being thrown forwards, the knee must be bent forwards, the thigh backwards, and the spine forwards again, while the free, natural walk which constitutes one of the chief beauties of woman is degraded into a tottering hobble. A walk of a mile in such boots is almost impossible, and if it were accomplished, would fatigue the wearer more than ten miles in boots or shoes which were made in accordance with the structure of the foot.

The muscles of the legs have their work completely changed, and so they become wearied, while the tendons are strained and the joints swollen.

---

Now for the battle of the shoes.

Begun no one knows when, the date of its final settlement seems equally obscure. Putting aside the Japanese straw shoe, and the American Indians' 'parflêche' shoe, we come to the shoe of iron, attached to the hoof by nails. The most curious point in this warfare is, that the more carefully the different inventors study the structure of the hoof, the more ingeniously do they contrive to inflict damage where they are really desirous of doing good.

I again ask the reader to disabuse his mind of

the idea that I bring any accusation of intentional cruelty against those who are entrusted with the care of the horse's hoof. On the contrary, I am sure that with very few exceptions, and those no more than the inevitable 'black sheep' which infest every business and profession, the farriers, grooms, and stablemen are really actuated with the best intentions towards the horse, and simply follow the traditions in which they have been brought up.

Having in mind the praiseworthy object of benefiting the horse, each inventor has recognised the defects of existing shoes, and has tried to produce a shoe which shall be free from these defects. The number of such shoes is so great, that I can only mention one or two of the most prominent.

Some years ago Mr. Goodenough, an American, recognised the fact that the broad flat surface of the shoe soon becomes so polished by friction that the horse could with difficulty gain a purchase on smooth ground, even if it were dry and level, while on sloping and wet ground the horse was nearly certain to fall. So he invented a shoe which was scooped beneath so as to present several sharp edges to the ground instead of a continuous smooth surface.

Here, then, was a recognition of a defect, and an attempt to remedy it. The shoe attracted much attention not only in America but in England, and I have seen many references to it as the only shoe which gave the horse a grip of the ground.

Certainly, for a short time this was the case. But the inventor had forgotten the rapidity with which an iron shoe is ground down by friction upon a hard road, and the result was, that the projections soon wore away, and the Goodenough shoe became just as smooth and slippery as any of the shoes which it was intended to supersede.

It had its period of popularity, but has long ago passed into the limbo of shoes which have been tried and found wanting.

Then came a time when the expansive power of the hoof was brought prominently forward, not to say exaggerated. It was recognised that the existing shoes, being unyielding in their structure, prevented the incessant expansion and contraction on which the health of the hoof largely depends, and so a number of shoes were devised which should permit the expansion and contraction. I will briefly describe three of them.

The first is the Clark jointed shoe, to which a casual reference has already been made.

This shoe is formed of two parts, each occupying one side of the hoof, and uniting at the toe, where they are joined by a rivet which allows the two halves to move freely. On paper and as a theory, the jointed shoe seemed as nearly perfect as anything could be, and it earned considerable praise. But Mr. Clark had overlooked three very important defects. The first was, that the footsteps of the horse had the same effect upon the rivet as so many blows of a hammer, fixing it so tightly that the sides lost their play.

The second defect was, that the iron of the toe always wears away faster than at any other portion of the shoe. Now, as the rivet was in the toe, the natural result was, that it soon became so much worn, that it fell out altogether, leaving the two loose ends as levers by which the nails were loosened at every step.

The third defect lay in imperfect knowledge of the structure of the hoof. The reader may remember that the lateral expansive property rests in the quarters, and that the toe is practically non-expansile. Therefore, as the shoe was nailed equally round the hoof, the quarters could not expand when fastened to the toe, and so there might just as well have been no joint at all.

In writing of this shoe, Lieut. Douglas, in the gently sarcastic tone which pervades his book, remarks that if the nails were let into half-inch slots, so as to have a quarter of an inch play in either direction, 'the jointed shoe would have some chance; as it is, there is none for it whatever.'

Another inventor realised the latter defect, and so the celebrated screw-shoe was devised.

This was a most complicated and expensive piece of mechanism, consisting of four distinct parts. First came the toe-piece. This was something like the capital letter T, with a curved cross-piece. The stem passed completely under the middle of the hoof, and had at the end a nut which worked the screw after which the shoe was named.

To the arms of the toe-piece the side-pieces were connected by rivets similar to those of the jointed shoe, their other ends or 'heels' coming against the screw. So, by turning the nut, the sides could be pushed further apart, or brought nearer together, so as to suit either a round or a long hoof.

This would have been a delightful arrangement if, like its predecessor, the screw-shoe could have remained on paper. When reduced to practice, however, its failure was even more disastrous.

In the first place, there were two rivets instead of one, so that it was doubly weak. In the next, the screws, which were really superfluous appendages, caused the stem of the toe-piece to work up and down at each step, until they fell out and were lost. But the third defect was an absolutely ludicrous one. The reader will remember that the expansion and contraction of the hoof is produced by the pressure of the frog upon the ground. Now, the stem of the toe-piece passes *under* the frog, and consequently prevents it from touching the ground.

A fair idea of the appearance of the screw-shoe may be obtained by comparing it to the ecclesiastic capital letter ℳ, the central perpendicular line representing the stem of the toe-piece.

The special object of the screw-shoe was, however, to be a corrective of contraction of the hoof, one of the ailments to which that much-enduring organ is subject. What is the practical working of the screw is well pointed out by Lieut. Douglas:—

'The foot was kept in water *to soften the horn* (!), and every day the screw was slightly turned by a key, the hoof apparently opening wider at the heels by the interior lengthening of the screw. But it was soon found that no proper expansion took place, it being either the nails that gave or the crust that split.

'So ended the theory of the screw-shoes, which could never have been looked into properly, as they would not have stood the test of common-sense investigation an hour.

'We have only to imagine that if the nails did not give, nor the crust split while the shoe was forced wider apart, this result must follow—viz. the wall of the foot would be torn away at the heels from the sole and the laminæ, a much more serious consequence than contraction could ever have produced.'

The reader will not fail to notice that whereas the hoof was intended by Nature to be nearly as hard as iron, man, in his attempts to improve upon Nature, does all in his power to soften it.

The third device for permitting the expansion of the hoof is the unilateral system, which has already been briefly mentioned. Provided that the frog be allowed to rest on the ground, and the shoe be not carried round the toe, as it is in some instances, this plan is fairly successful with the unshod side of the hoof. But on the other side, as the expansile quarter is connected with the non-expansile toe by a bar of rigid iron, it is rendered immovable, and the value of the frog is half lost.

Still, though the unilateral system is certainly better than the employment of a shoe which runs all round the hoof, it cannot but have an injurious effect on the hoof. That one side of the hoof should be in its natural elastic condition, and the other should be shod with unyielding iron, is necessarily as annoying to a horse as it would be to a human being.

If the reader has followed my train of argument he will see that if a shoe of any kind is to be nailed upon the hoof, it ought to be confined to the front portion, which we call the toe, there being no expansive properties to be neutralised by the shoe.

One great gain will be evident, namely, the lessening of the weight which the horse has to lift at each step. Also, only half the usual number of nails will be required, so that the horn of the wall is not so much weakened by nail holes. There is also no interference with the frog, which can assume its normal size and form, so as to take the weight of the body before the toe comes into use. I have already mentioned that the importance of the frog is rapidly becoming known in America. Mr. J. W. Gerry, in a letter to the 'Boston Globe,' puts the point very quaintly :—

'Customers say "My horse needs shoeing; his frog comes to the ground, and he begins to go lame." What an absurd idea, when Nature intended him to travel on the frog! *If not, why was not the frog put upon the top of the hoof?* Nine-tenths of all driving horses have corns because of the iron thick-heeled shoes to keep the frog from the ground.'

One thing yet is wanting.

The tip is fastened *upon* the hoof, and is therefore liable to be wrenched off if the horse should strike its toe against a hard substance. Even if it be only loosened, it becomes dangerous, as the nails are apt to be partly drawn and twisted out of the right direction. As the horse goes on trotting or even walking, the nails become more and more bent, and their points are liable to penetrate the linea alba, an accident which often results in lockjaw. In Mayhew's well-known work on 'Horse Management,' p. 83, there is a very good figure of the mischief which may be wrought by a loosened shoe.

Even granting that no such accident happen, and that the shoe retains its position, the horn of the toe is kept off the ground, and so cannot be subject to the friction which is necessary for its benefit. It therefore continues to grow unchecked, while that of the quarters is incessantly worn away and renewed as it ought to be.

Now, granting for the sake of argument that the horse *must* have a shoe, it is evident that the best shoe is that which interferes least with the natural growth and action of the hoof. Such a shoe is the 'Charlier,' so named after its inventor. It is extremely narrow, and very light; but its essential excellence lies in the fact that it is not nailed *upon* the hoof, but is countersunk into the horn, so as to be flush with the natural edge.

The Charlier shoe is made of various sizes. Most of them are 'full' or 'three-quarter' shoes, as seen in the accompanying illustration. These, however, are not to be recommended, as they are apt to work loose towards the heels, and interfere with the proper development of the quarters. As the reader will see, the heel and frog are left alone, so that the latter can grow to its normal size, nearly filling the cavity of the hoof.

A 'quarter' shoe is all that is required. There is a specimen now before me, which was taken from a large hoof where it was doing good work. When first put on, or rather put in, it weighed exactly three ounces, and has lost a quarter of an ounce by wear. It is three-sixteenths of an inch in thickness and seven-sixteenths in width. If it were straightened it would measure rather more than seven inches in length. Now, this is altogether too large a shoe, but I describe it as being an actual specimen which had been in use.

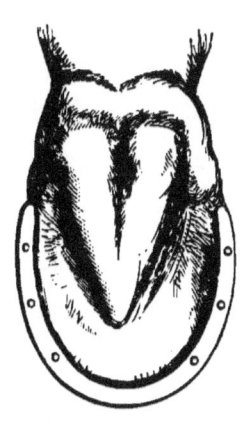

CHARLIER SHOE.

Seven inches, for example, is much too long, four inches being amply sufficient for practical purposes. It is also too thick, one-eighth of an inch being quite enough; and it is needlessly wide, the few nails which are required being so small that there is plenty of room for them if the width were diminished by at least one-fourth.

A little art is required in attaching this shoe. Round the edge of the hoof a groove is cut, and into the groove the shoe is sunk. The groove should be fully as deep as the thickness of the shoe, and even should it be a little deeper no harm will be done. There is no difficulty in cutting the groove, as in all places where the Charlier shoe is sold a special knife is supplied. This knife is a variation of the ordinary draw-knife, and is furnished with a moveable guide by which the dimensions of the groove are regulated. The guide can be set to any point, and fastened by a screw, so that even the most careless operator can hardly make a mistake.

As the shoe is flush with the horn, there is no leverage, and in consequence it only requires a few nails, and those of small size. The chief virtue of the Charlier shoe, however, consists in the fact that, as the horse steps, the horn and iron are worn away simultaneously, so that the shoe can be worn down until it is hardly thicker than a visiting card.

It is not very easy to induce farriers to fix the Charlier shoe to the hoof. All artisans have a tendency to move in ruts, and are almost helpless when taken out of them, and the farrier is perhaps more wedded to his accustomed rut than any other workman.

Should the full, or three-quarter, Charlier be

employed, it is always easy to cut away a piece from the heels at each time of renewal, so that in three months or so nothing will be used except the four-inch tip. No matter how much a hoof may have been mangled by ignorant men, it will at the end of that time have regained nearly all its original form.

I am aware that in one of our cavalry regiments the Charlier shoe was tried and failed, on account of the shoes breaking. But the shoes were far too large, coming back as far as the heels, and of course being liable to twisting and breaking. Had they been four-inch tips, a very different result would have been obtained. Moreover, it is clear that the farriers were not desirous to have a shoe which they did not understand.

WE are now brought face to face with another development of the subject. It has been shown that the horse's hoof, when untouched by man, is one of the most perfect and elaborate pieces of animal mechanism in the world. Common sense ought to tell us that when man meddles with Nature he always muddles, and that with the best intentions in the world he deliberately sets himself to undo Nature's work.

In nothing does man seem to revel in his opposition to Nature so much as in his dealing with the horse's hoof. He cannot let it alone, but cuts, and carves, and scoops, and rasps, and greases, and blacks, and rends its fibres as if he were not guided by reason, but by a series of 'happy thoughts' which happened to occur to him, and were immediately acted upon without the least reflection.

For example, the Creator has taken the greatest care to make the whole Hoof as light as possible. 'Happy thought!' says man. 'Let us hang a pound or so on each hoof and make the horse waste his strength in lifting it.'

He has made the Wall exceedingly strong. 'Happy thought! Let us weaken it by cutting it away, by scooping grooves in it, and driving nails into it so as to tear the fibres asunder.'

He has made this wall nearly as hard as iron. 'Happy thought! Let us soften it by "stopping" and similar devices.'

The Creator has made the edge of the wall quite sharp, so as to enable it to hitch upon the slightest unevenness, and to aid it in ascending a hill. Another happy thought! 'Let us cut away the sharp edge, and substitute a flat, smooth surface of iron which can take no hold of slippery ground.'

There is now before me a shoe taken from the foot of a dray-horse. The flat iron surface is exactly two inches in width, so that if man had deliberately set himself to work to invent a plan of making the horse's footing as insecure as possible, he could not have been more successful.

He has furnished the hoof with an elastic pad called the 'Frog,' so as to prevent any jar when the horse steps. 'Happy thought! Let us cut away the pad and make the horse's weight come upon a ring of iron.'

This happy thought has another effect. The Creator has constructed the hoof so that the elaborate and delicate mechanism of its interior can only be kept in working order by the pressure of the pad, or 'frog.' So by cutting away the frog in one of his happy thoughts, man has contrived to stop all this machinery (with the inevitable result to machinery which is not allowed to work), and has also managed to create exactly the jar which the frog would have prevented.

A third object has been attained by cutting away the frog. As has already been shown, the frog prevents the horse from falling on smooth and slippery surfaces, and even enables it to gallop over ice. So, by removing the frog, the chances of a horse's falling are greatly multiplied.

Again, the Sole of the hoof has been formed arch-wise, of successive layers of exceedingly hard horn, so as to perform a double office. In the first place it bids defiance to hard and sharp-edged objects, such as rocks or broken flints; and in the next place, it transmits the pressure from the frog to the wall, so as to produce the expansion at each step on which the health of the hoof depends.

So the sole inspires man with another happy thought. 'Let us pare it so thin that it not only cannot resist the pressure of the horse's weight upon a stone, but that it yields to the pressure of the human thumb. As for its duty of transmitting pressure from the frog to the wall, that is of no consequence, as the frog has already been cut away, so that there is no pressure to be transmitted.'

Our happy thoughts are not yet exhausted. The hoof has been made Porous, so as to permit transpiration to take place. 'Happy thought! Let us choke up all the pores with oil to stop the transpiration.'

It has been covered with a natural Varnish so constructed that while it does not hinder transpiration, it defends the hoof from wet. 'Happy thought! Let us rasp or scrape off the varnish. It is natural, and therefore must be wrong.'

The coronary ring, from which the fibres of the wall are secreted, is guarded by a Penthouse of Hair which causes wet to shoot off it as it does from the eaves of a house. 'Happy thought! Let us snip away the hair, and let the water make its way into the coronary ring.'

So, after working his sweet will upon the hoof, man wonders at its weakness, and lays down the stupid axiom that 'one horse can wear out four sets of legs,' which is equivalent to saying that the Creator did not know how to make a horse.

## CHAPTER IX.

The shoe useful in proportion to its lightness—Therefore, the best shoe seems to be none at all—Capability of the human foot—Value of an army dependent on its marching power—Lord Wolseley's axiom—Edinburgh lasses—Moccasin *versus* boot—Mansfield Parkyns in Abyssinia—Ladies and children at the sea-side—Charles Waterton in Guiana—Col. Dodge's account of the North American Indian's pony—A race between the Indian's pony and the high-bred horse—Exmoor and Dartmoor ponies—Description of these 'moors'—How to make a horse's hoof tender—The hoof an organ of all-work—Saddle and draught—Col. Burnaby's opinion.

Now we are brought face to face with a problem which cannot well be avoided. We have seen that the value of the shoe is in inverse ratio to its weight, its dimensions, and its interference with the natural functions of the hoof.

The thick, heavy shoe, with calkins, clips, and toe-piece, is the worst, while the light Charlier tip, which is countersunk into the horn of the toe, is the best. The logical conclusion is, that as the horse is benefited in proportion as the shoe is diminished, it might, unless under very exceptional circumstances, be more benefited by having no shoe at all.

No one who has the least acquaintance with the

horse's hoof can deny that every one of the numerous ills which beset a horse's hoof is caused by the iron shoe and its adjuncts—namely, the knife, the rasp, the nail, the 'oils,' and the stopping. Let us impartially compare the advantages and disadvantages, and see on which side the balance inclines.

We may take ourselves as examples.

We, who have been accustomed through all our lives to have our feet defended by boots and shoes, would be lamed and bleeding in a few minutes if our artificial protections were removed, and we were obliged to run or even to walk for half a mile on hard ground. Yet the human foot was not intended by the Creator as an instrument for maintaining bootmakers, and is perfectly capable of bearing its owner over ground which would cut the best made boots to pieces.

Military life, in which the soldier has to walk for considerable distances daily over all kinds of ground, affords a good test of the powers of the human foot. Next to the commissariat, which feeds the men, scarcely any department causes such anxiety as that which deals with the feet on which the army is conveyed from place to place. 'The army that can march best is the best army,' writes Lord Wolseley; 'and the regiment that can march best in an army, is the best in that army.'—('Soldier's Pocket Book,'

p. 257.) The care of the soldier's foot is repeatedly urged in that valuable work.

Yet, with all the care that can be taken, men who are obliged to wear boots and shoes are no match in marching for those who have always been accustomed to go barefooted. Most of the disasters that have befallen our troops when dealing with uncivilised races, such as the Maori, the Zulu, and the Arab, have been due to ignorance of the tremendous marching powers of the barefooted savage.

I never fairly appreciated the capabilities of the human foot until February 1881, when I was staying in Edinburgh. It was a very severe winter, having been made memorable by the terrible snowstorm of January 18.

The Cowgate (locally pronounced Coog't) is now mostly inhabited by the poorer classes, the children of both sexes and the younger women habitually going without shoes. I was obliged repeatedly to cross this street, which was then in extremely bad condition. Mud had been deep, and the cartwheels and horses' hoofs had cut it up into a very rough state. Then a severe frost had come on, and the result was that the frozen ridges were so hard and sharp that they hurt my feet through the thick soles of my boots.

Yet these children and young women ran and

skipped over the road as freely as if it had been soft turf, and suffered no inconvenience from the sharp surfaces. Still more curious was the way in which they stood and talked unconcernedly, though their naked feet rested on ice and snow, and their legs were bare half way to the knee. I asked several of them whether they did not suffer from the cold, but they all agreed in saying that they did not feel the cold inconvenient, except when they wore boots.

It does not need that boots or shoes should never have been worn, in order to produce this insensibility of foot. Hunters in North America always abandon the boot for the moccasin, which is only a single thickness of hide lashed over the foot. At first, the civilised hunter finds walking very painful, but in a short time his feet become accustomed to their work, and the man almost looks forward with horror to the time when he must return to civilisation and boots.

When Mr. Mansfield Parkyns was on his hunting tour in Abyssinia he very wisely conformed to the dress of the natives, and always went barefooted. He was soon able, as they were, to follow the chase on foot, over sand, or rocks, or through bush. After the day's hunting was over, he, like the natives, had his feet overhauled, in order to find whether any thorns or splinters might be sticking in them.

On one occasion he narrates how he incautiously

trod on an upright and broken stump, and ran a splinter so deeply into the sole of his foot that the point could just be seen as a dark spot under the skin of the instep. He could not afford to stop, so he continued the chase, killed his quarry, and carried it home, a distance of some two miles. He then cut down upon the broken splinter, so as to expose the end, and pulled it out with his bullet-mould. In three or four days his feet were all right again.

The late Charles Waterton, in his 'Wanderings in South America,' invariably went barefooted, just as the natives did.

The rapidity with which Nature will enable the tender feet of civilised man to be as tough as those of the savage is really wonderful. A year or two ago there was a fashion for families to go to the seaside, and walk about all day without shoes. At first the experience was rather painful, even upon tolerably smooth ground. But, before the time of the visit had nearly expired, ladies and children could be seen tramping on the shingle, or clambering over sharp-edged rocks, with perfect ease, though at first each step cost a separate pang.

Quite as remarkable is the completeness with which the toughness of the foot disappears when it is not required. Mr. Waterton afforded a curious instance of this adaptability. He proposed to a friend to walk to Rome from Baccano, and, in order to be more comfortable, he determined to walk barefooted.

'Having,' he writes, 'been accustomed to go without shoes month after month in the rugged forests of Guiana, I took it for granted that I could do the same on the pavement of his Holiness Pope Gregory the Sixteenth, never once reflecting that some fifteen years had elapsed from the time that I could go barefooted with comfort and impunity. During the interval, however, the sequel will show that the soles of my feet had undergone considerable alteration.'

The result to which he alludes was that his feet were terribly cut and bleeding before he had traversed four miles, and that he was confined to the sofa for two months. Of course the pavement was comparatively smooth, and not nearly so likely to injure the foot as the roadless ways which he traversed in Guiana, but the sole of the foot, for want of use, had lost its former toughness.

I use this last word intentionally. The popular idea, which until within a few years I myself shared, is that when the boot, or shoe, or sandal is not worn, the sole of the foot becomes hard. Now, so far from being hard, it is quite soft. It is true that the skin becomes extremely thick, but at the same time it loses none of its flexibility, and, like the frog of the horse, or the pad of a dog or cat's paw, has very much the consistency of indiarubber, so that sharp stones, thorns, &c., have but very little effect on it.

We will now return to the horse, and see what can be done by a horse which has never known a shoe. Here is an account by Lieut.-Col. R. J. Dodge, U.S.A., of the horse as used by the red men. These horses are the descendants of the animals which were brought over by the Spaniards and afterwards abandoned. Being left to themselves, they multiplied exceedingly, and reverted to the wild state, forming themselves into herds, whence they are captured when wanted by means of the lasso:—

'My subject would not be complete without some mention of the Pony, the plain Indian's inseparable companion and most serviceable slave.

'Scarcely fourteen hands in height, he is rather light than heavy in build, with good legs, straight shoulders, short, strong back, and full barrel. He has no appearance of "blood" except sharp, nervous ears and bright intelligent eyes; but the amount of work he can do, the distance he can pass in a specified time, provided that it be long enough, put him in Indian hands fairly on a level with the Arabian.

'Though of indispensable value to the Indian, he receives not a particle of attention. He is never stabled, nor washed, nor rubbed, nor curried, nor blanketed, nor shod, nor fed, nor doctored.

'When travelling over rough and rocky ground, his rider *may* take the trouble to tie up a tender foot in a piece of buffalo robe.

'After endurance, the best quality of the pony is sureness of foot. He will climb a steep rocky hill with the activity and assurance of the mule. He will plunge down an almost precipitous declivity with the indifference of the buffalo.

'For going over swamps and marshy places he is only excelled by the elk, and he will go at speed through sandhills, or ground perforated with gopher holes, where an American horse would labour to get along at a walk, and fall in the first fifty yards of a gallop.

'The amount of work got out of him by the Indian is astonishing. No mercy is shown. Tell an Indian to find out something miles away, and he will probably go and return at full speed, though the distance made be twenty miles. And this work is done under apparently most unfavourable circumstances: a terrible bit, an ill-fitting saddle, and a rider as cruel and remorseless as Fate itself.'—('Hunting Grounds of the Great West.')

It is scarcely possible to imagine a stronger testimony to the endurance and sure-footedness of the unshod horse. That speed should be a characteristic of such an animal could not reasonably be expected. At all events, no one would suppose that the pony so nurtured could equal in that respect the pedigree horses of civilised man, who not only selects the parents from families renowned for speed and power, but carefully trains the offspring for racing.

Yet, that an animal so neglected according to our ideas as the Indian pony can not only equal but surpass the trained horse of the white man, is shown by a most amusing account of a race between a pony and blood horses, written by Lieut.-Col. R. J. Dodge in another part of his valuable work :—

'It is exceedingly difficult to hit on a fair distance between the Indian and American horse. The start being always from a halt, the small, quick pony is almost sure to win at from one to three hundred yards, while the long stride of the American horse is equally sure of carrying him in winner at from six hundred yards to two miles. A mile or two is then doubtful, after which it is safe to back the endurance of the pony.

'A band of Comanches under Mu-la-que-top once camped near Fort Chadbourne, in Texas, and were frequent visitors and great nuisances as beggars at that post. Some of the officers were decidedly "horsey," several owning blood horses, the relative speed of each being known by separate trials almost to a foot. Mu-la-que-top was bantered for a race, and after several days of manœuvring, a race was made against the third best horse of the garrison, distance four hundred yards.

'The Indians betted robes and "plunder" of various kinds, to the value of sixty or seventy dollars, against money, flour, sugar, &c., to a like amount. The Indians "showed" a miserable sheep of a pony, with legs like churns, a three-inch coat of rough hair stuck out all over the body, and a general expression of neglect, helplessness, and patient suffering which struck pity into the hearts of all beholders.

'The rider was a stalwart Indian of one hundred and seventy pounds, looking big and strong enough to carry the poor beast on his shoulders. He was armed with a huge club, with which, after the word was given, he belaboured the miserable animal from start to finish. To the astonishment of all the whites, the Indian won by a neck.

'Another race was proposed by the officers, and after much "dickering" accepted by the Indians against the next best horse of the garrison. The bets were doubled, and in less than an hour the second race was won by the same pony, with the same apparent exertion and with exactly the same result.

'The officers, thoroughly disgusted, proposed a third race, and brought to the ground a magnificent Kentucky mare of the true Lexington blood, and known to beat the best of the others at least forty yards in four hundred. The Indians accepted the race, and not only doubled the bets as before, but piled up everything that they could raise, seemingly almost crazed with the excitement of their previous success.

'The riders mounted, the word was given. Throwing away his club, the Indian rider gave a whoop, at which the sheep-like pony pricked up his ears and went away like the wind, almost two feet to the mare's one. The last fifty yards of the course were run by the pony with the rider sitting face to the tail, making hideous grimaces, and beckoning to the rider of the mare to come on.'

The reader will probably have noticed the enormous weight of the man who acted as jockey, as well as the fact that when the rider really wanted the horse to do its best he abandoned the stick and only urged the animal with his voice.

The hardness and sure-footedness of the unshod horse is not confined to the American animal, and neither are due to the effects of climate, as is urged by many objectors. Here, for example, is an account of the wild Exmoor ponies, which, in many respects, resemble those of the American prairies. It is taken from Sidney's 'Book of the Horse,' a work which by no means advocates the theory that horses, as a rule, need no shoes:—

'Horses bred on the moors, if left to themselves, rapidly pick their way through pools and bogs, and canter smoothly over dry flats of natural meadow, creep safely down precipitous descents, and climb with scarcely a puff of distress these steep ascents; splash without a moment's hesitation through fords in the forest streams swelled by rain, and trot along without a stumble along sheep-paths bestrewed with loose stones.

'A sight scarcely less interesting than the deer was afforded by a white pony mare with her young stock, consisting of a foal still sucking, a yearling, and a two-year old. The two-year-old had strayed away feeding, until alarmed by the cracking of our whips and the neighing of its dam, when it came galloping down a steep coombe, neighing loudly, at headlong speed. It is thus that these ponies learn their action and sure-footedness. One of these little animals, barely four feet (12 hands) in height, leaped standing over a barrier five feet high, barely touching it with his hind feet.'

In order to realise the force of this description, the reader must first understand what kind of places are Exmoor and Dartmoor. Most people take for granted that moors and prairies are level surfaces of great extent, covered with grass, and as easy to the feet as if they were paddocks. Mr. G. Ransom, in his 'Horses and Roads,' quotes the account of a Devonian :—

'Dartmoor is not a great wild flat, as many suppose, but, on the contrary, it is for the most part a continual succession of very steep, rough hills, or 'tors,' and rugged 'coombes,' strewn with granite rock and stones. Yet, in spite of all, besides the bogs and *chronic state of rain*, the herds of ponies gallop fearlessly along the rough sides of the coombes, down and up. It is a pretty sight to see them, especially in the spring, with the foals by their sides.'

Another writer, Lieut. W. Douglas, who has been frequently quoted in these pages, is equally strong on the subject :—

'From the moment a horse is foaled, we either keep him in fields soft to tread upon, or in warm stables, standing on soft straw, and then we are surprised that his hoofs should become dry and brittle, instead of keeping moist, tough, and *hard*.

'In the Orkneys, in the mountains of Wales, the wilds of Exmoor and Dartmoor, in many parts of the continent of Europe, and for a considerable portion of the rest of the globe, horses run about over rocks, through ravines, and up precipitous ridges, unshod, and this to the evident advantage of their hoofs, for these animals never suffer from contracted feet, or from corns, sandcracks, &c., *until they have become civilised and been shod.*' (Compare Lady Florence Dixie's account, p. 25.)

It may probably be objected that the ponies of Exmoor and the wild horses or 'mustangs' of America have neither burdens to bear nor loads to draw. We will meet both objections.

We have just seen that the mustangs are ridden by very heavy riders for long distances, and over the hardest and roughest ground. We have also seen that when the Indian rider trains his horses, he never dreams of putting iron shoes on its hoofs, and that the horse does this exceptionally hard work without requiring shoes.

Now I find that there is almost an invariable point among objectors. They admit that horses might get on without shoes under different circumstances from those in which they happen to be placed. For example, those who live in low-lying and moist districts, such as are found in the Fen countries, say that horses might possibly do without shoes on hard and hilly countries, whereas their hoofs would be ruined by being always in contact with the roads that pass through the damp soil of marsh land.

Then, those who inhabit the hard hilly countries are equally sure to say that horses may travel unshod on a level and moist surface, but that their hoofs would be cut to pieces on their own stony hills.

When they are 'driven to their last ditch' by being confronted with a horse which is perpetually ridden and driven over the worst roads in England, and which has worn no shoes for many years, they say that the animal is an exceptional one. The same series of opposition has to be encountered by anyone who dares to think that the Creator can make a hoof which can do any kind of work, in any climate, so long as it is let alone.

Generally, I find that although many people are ready to admit that unshod horses might answer for riding purposes, they cannot believe that the unprotected hoof will permit the animal to draw a load behind it. I was not prepared to hear a precisely opposite opinion, and that from a man whose judgment is worthy of all respect. I mean the late Lieut.-Col. Fred. Burnaby, author of the 'Ride to Khiva.'

In the course of a conversation on the subject, he stated that he considered the horse as capable of drawing loads without shoes, but not of bearing a rider of ordinary weight, say eleven stone. The additional weight would, in his opinion, wear away the hoof so fast that Nature could not renew it sufficiently. As to cavalry, where the horse has to carry an abnormal weight, *i.e.* never less than eighteen stone, he was sure that the hoof could not endure the work. He expressed the same opinion to Mr. A. F. Astley, in a letter which I possess.

One person who was questioned on the subject gave a precisely opposite opinion. I cannot quite follow the argument, but insert it in hopes that someone else may be more fortunate. 'As driving does not tire a man, and riding does tire a man, *therefore* an unshod horse if driven would have a harder time of it than if ridden!'

I hope to show not only that a horse *may* do either saddle or traction work without shoes, but that he actually *does* both kinds of work on any description of road, and better than when he was shod. The reader will observe that I put forward no theories, but state facts, and produce proofs.

## CHAPTER X.

Unshod horses now at work—Dr. Llewellyn's horse on London roads—Thirteen thousand miles without shoes—Always went lame when shod—His transfer to Mr. A. F. Astley—Mr. Astley's horse 'Tommy'—State of his hoofs when bought—Process of training—Work done by him unshod on hard roads—Photographs of his hoofs—Mr. Whitmore Baker's mare 'Stella'—Facts *versus* theory—Photographs of 'Stella' and her hoofs—Letters from Mr. Baker—Work done by 'Stella' barefooted—Galloping over ice or loose stones—Mr. Baker's offer to enable other horses to work unshod—His preparation for hoofs and its possible value—Influence of external conditions on the hoof—The condemned tramcar horse—Result of removing the shoes—Five hundred and forty miles unshod—Photograph of the hoof in transitional state—Mistaken benevolence.

ARE there any living horses which can work without shoes? If so, how was this condition of hoof attained? How long is it likely to last, and on what roads did the horse work?

As to the first question, I can produce the experience of many owners of shoeless horses, but prefer to take one or two as examples, because I am empowered to give their names.

One is a very remarkable horse which was driven unshod for several years by Dr. Llewellyn, of London, and which afterwards passed into the possession of my lamented friend, the late A. F. Astley, of 46 Great Cumberland Place, Marble Arch, W. Here is a copy of the certificate signed by the former owner :—

'I certify that for five and a half years the horse (whose hoof is here photographed) has done my work (a doctor's) BAREFOOT—that is to say, without iron of any kind on his feet.

'This work he has done single-handed, for I keep but one horse. Most of his work has been done over macadamised roads in the East of London.

'He has often worked seven days a week, and has frequently had a heavy brougham behind him. In the five and a half years he must have traversed some 13,000 miles. Daily work, three or four hours. Though barefoot, he has worked sound, and his hoofs show no signs of undue wear.

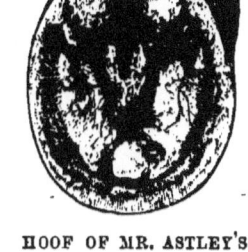

HOOF OF MR. ASTLEY'S HORSE, PURCHASED FROM DR. LLEWELLYN.

(Signed) 'R. RALPH LLEWELLYN.
'July 20, 1883.'

I here present to my readers an engraving of a photographic portrait of the near fore-hoof of this horse. Let the reader, when examining it, bear in mind that the horse had traversed at least thirteen thousand miles without any protection to the hoof. Another remarkable point in the history of the animal is, that he always went lame when shod, but has never done so since his shoes were removed.

In order to give the reader a more complete idea of the appearance of the hoof when untrammelled by the shoe, I here present him with a fac-simile of a photograph taken from the two fore-feet of the same animal.

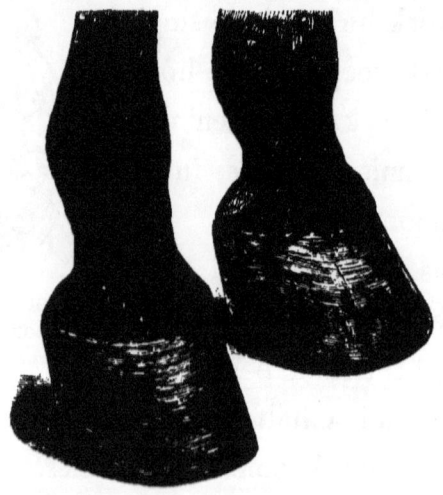

FORE-FEET OF DR. LLEWELLYN'S HORSE.

I need only request my unprejudiced reader to look at this photograph, and ask himself whether any shod hoofs can compare with these. I do not offer a fancy drawing of hoofs, as I think they ought to be under certain circumstances, but give the photographic portrait of hoofs as they *are* under those same conditions.

I may here mention that when the photograph was taken the horse was fourteen years old, and that its height was fifteen hands one and a half inches.

Here is another photograph, representing the hoof of 'Tommy,' another of Mr. A. F. Astley's horses.

I possess many letters from Mr. Astley in which he describes the trouble which he had with 'Tommy,' who was about as unsatisfactory a subject for an experiment as could well be imagined, having only one fairly sound hoof out of the four. Several times his owner was on the point of abandoning the enterprise as a failure, so fragile was the hoof. Perseverance, however, as mostly is the case, succeeded at last, and 'Tommy' was able to work without shoes far better than he had done while wearing them.

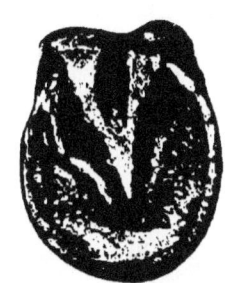

HOOF OF MR. ASTLEY'S HORSE 'TOMMY.'

The history of 'Tommy' is briefly as follows.

When the photograph was taken, the horse was sufficiently old for his age to be uncertain. His colour was chestnut, and his low-heeled hoofs white. He had then been constantly worked for ten months without shoes or any protection for the hoof.

The day before the photograph was taken, Mr.

Astley drove 'Tommy' about twenty-four miles on hard road.

The two longest drives that had then been taken were these. The first from London to Watford and back, between twenty-nine and thirty miles. This was taken on March 29, 1883. On May 14, of the same year, Mr. Astley drove 'Tommy' to St. Albans and back—*i.e.* about forty-two miles.

During this process, Mr. Astley sent me a series of bulletins, sometimes in letters, which at first were quite despondent, and sometimes by cards, stating the number of miles which 'Tommy' had traversed.

If the reader will contrast the photographs of these two animals, he will see that 'Tommy's' hoof has scarcely any concavity, and therefore is not quite so perfect an example as that of the former animal. Yet, although not a perfect hoof, it is a good and sound one. The bars are boldly marked, and the frog has become so largely developed, that it fills up almost the whole of the concavity. With such a hoof as this, the animal need not fear the pointed tips of shattered rocks or the razor-like edges of broken flints, and would gallop over either in serene unconsciousness of their existence.

Almost identical with this hoof is that of a mare who has already become historical in the Battle of the Shoes, and around whom long raged a wordy battle, if it can be called a wordy battle, where the words were all on one side and the facts on the other. On reading this controversy, I am irresistibly reminded of one in which I was an unwilling partaker.

Like many better men than myself, I had been misled by popular prejudice, and had thought the bulldog to be a stupid, fierce, morose, savage animal, always ready to assault any being except his master, and good for nothing except fighting.

By some chance it happened that I became the possessor of an infantile bull pup, and in consequence had to alter all my ideas on the subject. I found that 'Apollo,' as I named him, because I then lived at Belvedere, in Kent, was the sweetest in temper, the most loyally affectionate, and, with one exception, the most intellectual dog that I ever knew. No bloodhound could surpass him in scent, and no retriever could have beaten him in water or on land.

As in duty bound, I took the first opportunity of recanting my former opinions, and was very much amused by the criticisms to which I was subjected.

On the one side, a leading journal in the sporting interest was derisively incredulous, on the ground that no bulldog could have behaved in such a manner, therefore that 'Apollo' could not have been a true bulldog, and that, therefore, I was unworthy of belief.

On the other side, the 'Saturday Review' made the following remarks, with which I entirely coincide :—' We cannot imagine how such a lover and observer of animals as Mr. Wood could ever, save in his greenest and most salad days, have entertained, as he confesses he did entertain, the vulgar and utterly unfounded notion that the bulldog is a savage and morose brute; the bulldog being as amiable an animal as walks the face of the earth.'

Very similar was the controversy respecting 'Stella,' the mare in question. She was of good lineage, being the daughter of ' Blair Athol,' and in 1880, when her owner, Mr. Whitmore Baker, then of Totnes, but now of Paignton, Devon, began his experiments, she was seven years old. She had been hunted in Devonshire for two seasons, and had been used as a hack and also for drawing a four-wheeled 'trap.' In December 1880 Mr. Baker had the shoes removed, and tried the experiment of allowing the animal to do her work upon her own feet, without a protection of any kind.

'I then rode the mare,' writes Mr. Baker, in a letter addressed to the 'Field,' newspaper, ' when snow and ice were on the ground last winter on the roads in this neighbourhood, and in the town of

Totnes. My weight being over twelve stone, and the roads almost impassable for horses shod in the ordinary way, I was highly gratified to find that the mare was able to trot over the slippery ground without fear of slipping, and this encouraged me to persevere with my treatment. After the nail-holes had disappeared in the natural growth of the hoof, the frog grew to its natural state, and the hoof assumed its natural form.

'From the time of my first experiment to this date (March 11, 1882) the mare has been without shoes, and has been worked both for riding and driving constantly on the roads in this neighbourhood, all of them hilly, and some of them very rough and trying to horses.

'The result has been that the mare has never fallen lame nor shirked work, but has stepped with greater freedom and ease than she ever did before. Sometimes, during this period of fourteen months, she has travelled ninety-six miles in one week, and now she is as sound as a bell, and never has suffered from the absence of shoes in any way whatever. Nay, more; before I removed the shoes she was occasionally lame, but since their removal she has not been lame once.

'Those of your readers who know what Devonshire roads are will, I think, concede that the test is a severe one. The result is, to my mind, conclusive, and the mare is with me now, unshod, and doing her usual work.'

Mr. Baker's success naturally drew upon him the wrath of those whose living depended on farriery, and their angry letters to the newspapers are numerous and amusing. Of course no real arguments could be produced, the writers having recourse to sarcasm, denial of facts, and invective.

One of these professionals, in a long letter to the 'North Devon Journal,' employs a most ludicrous series of objections. He cannot deny that the animal wears no shoes, and may be able 'to potter about, and do the little she has to do barefooted.' Then he states that the unshod hoof cannot withstand contact with the hard roads if the animal be worked regularly, entirely ignoring the fact that the horse *was* worked regularly, and had done nearly one hundred miles in a week.

Then he says that he has seen horses quite lame which had been working unshod on a farm, and therefore that it is impossible to work on a road without going lame. Then he accuses Mr. Baker of retrogression to the barbarian days when there were no blacksmiths, so that the horses could not obtain shoes, although they needed them.

Lastly, he takes a higher standpoint, and boldly

taxes Mr. Baker with irreligion. 'Did not He who formed the horse most *surely intend that we should live by one another*; and is it not civilisation that brought about industry of every description; and did not He who ordereth all things put it into the heart of man to study what could be done to protect the feet of that useful animal the horse?'

Which argument, being reduced to its elements, means that the Creator made the horse's hoof unable to do its work, so that farriers might gain a living by nailing iron shoes upon it. As I can scarcely expect any reader to believe that such fatuous nonsense could be put forward as arguments, I must refer him to the 'North Devon Journal' of May 11, 1882.

There is now before me a cast of the off-fore hoof of this very animal which was benighted and irreligious enough to do its work without shoes when the professional shoer said that it could not work unshod; or that, if it did, it was flying in the face of Providence. The cast was taken in December 1882— *i.e.* eight months after the letter was written.

In spite of the high lineage of the animal, the hoof is not a first-rate one, the slope being too great. From heel to heel the circumference of the wall is thirteen inches; from heel to toe it measures five and a quarter inches, and across the quarters it is a trifle more than four inches.

The weight of the horse was necessarily taken off the hoof when the cast was made. Had the opposite foot been lifted from the ground, so that the animal would have been obliged to rest upon the hoof, these dimensions would have been slightly altered.

I have exhibited this cast in most of the large cities and towns of this country, as well as in America, and in every case it has excited the greatest interest. The most remarkable point about it is the manner in which the under surface is filled up, the only cavity being that on either side of the frog there is a depression scarcely large enough to contain the last two joints of a lady's little finger.

Wishing to know whether in the autumn of 1884 'Stella' was still in the possession of hoofs as sound as those of 1882, and still did the same amount of hard work, I wrote to Mr. Whitmore Baker, and received the following answer:—

'I have the satisfaction of stating that my mare "Stella's" hoofs continue in the most perfect condition, and I shall be most happy if I can assist you. Devonshire may be an exceptional county (at least I hope so), for though my mare is known far and wide to travel shoeless, still, nothing further is said than, " It is wonderful ! " and " 'Stella' is one of a million."

' I have offered to take *any* horse sent to me, and to prepare the hoofs without any fee, simply the expense of keep being paid. But none have sufficient faith! Should any one you know be inclined to send me a horse, it would have my undivided attention, and be another test of the perfect success of my scheme.

'I must ask you to accept on my word the following advantages derived from my own experience :

' 1. Five or six pounds per annum are saved by non-shoeing, including the frost nails in winter.

' 2. I can gallop " Stella " on a road covered with ice, when other horses are not safe even with the use of frost nails.

' 3. The weight of the shoes is taken off the feet, which is a considerable help to the horse.

' 4. The foot, being flat from the frog, and down to the ground, leaves no receptacle for stones.

' 5. There is none of the unnecessary jar caused by the shoes, so that the horse travels freer and lighter.'

No question of theory can arise in this instance. The *fact* is, that the horse has been regularly working unshod over some of the most trying roads in England. The *theory* is, that no horse can perform such a task, or that, if so, the animal must be an exceptional one. Mr. Baker, resting on fact, generously offers to take any horse that may be sent to him, and to make its hoofs as sound as those of ' Stella.' Mr. A. F. Astley, also resting on facts, made his first experiment on an unsound horse; failed at first, but ultimately succeeded, and, but for his premature death, he would have picked out a series of horses condemned on account of their hopelessly foundered condition, and tried to give them another lease of life, together with the power of work and the capacity of enjoyment.

In another letter, dated Sept. 13, 1884, Mr. Whitmore Baker writes to me as follows:—

'There is one thing which I did not state in my last, and which will perhaps interest you. Having some knowledge of chemistry, I made a preparation for toughening horn; I then had my mare's shoes removed, the edges neatly rasped, and the nail-holes plugged up with whalebone. I then used my preparation three times daily, gradually exercising the mare over the road, and never allowing her to stand on straw except for bedding.

'Though I place great importance on the preparation, I do not say that a horse will not succeed without it. But I maintain that it facilitates the process by allowing the heels and frog to take a bearing before sufficient horn has been worn away to make the parts sensitive. I have known many instances where the shoes have been taken off and nothing done to the feet, and all have become lame within a month or so.'

Mr. Whitmore Baker's idea of plugging the nail-holes with whalebone is a singularly happy one, the whalebone and hoof being identical in structure, and the fibres running in the same direction. Ordinary horn would answer as well if cut longitudinally, but whalebone is much easier to handle.

With regard to the preparation invented by Mr. Baker, it is certainly not necessary, as is shown by the number of hoofs which have become strong and sound without it; but it only hastens the process of hardening, or adds to its chances of success.

The portrait of 'Stella' and her mistress (see Frontispiece) is engraved by permission from a photograph taken by Messrs. Brinley & Son, Fore Street, Totnes.

The owner of a soft-hoofed horse may naturally fear to expose the animal to such an ordeal as a journey upon hard and stony roads. Yet success is only a matter of time, the hoof having the power of accommodating itself to any kind of ground. This is forcibly shown by Professor Fleming. After mentioning the influence of climate, of weather, of health, of age, food, and labour upon the hoof, he proceeds as follows:—

'The seasons are to some extent concerned in the growth and shape of the hoof. In winter it widens, becomes softer, and grows but little. In summer it is condensed, becomes more rigid, concave, and resisting, is exposed to severe wear, and grows more rapidly.

This variation is a provision of Nature to enable the hoof to adapt itself to the altered condition it has to meet—hard horn to hard ground, soft horn to soft ground.

'In this way we can account for the influence of locality upon the shape of the foot. On hard dry ground the hoof is dense, tenacious, and somewhat small, with a concave sole, and a little but firm frog; while in marshy regions it is large and spreading, the horn is soft and easily destroyed by wear, the sole is thin and flat, and the frog is only an immense spongy mass, which is badly fitted to receive pressure from even hardened soil.

'In a dry climate we have an animal small, compact, wiry, and vigorous, travelling on a surface which demands a tenacious hoof, and not one to prevent sinking. In the marshy region, we have a large, heavy, lymphatic creature, one of whose primary requirements is a wide, flat foot to enable it to travel on a soft, yielding surface.

'Change the respective situations of these two horses, and Nature immediately begins to transform them and their feet. At first the light, excitable, vigorous horse, with its small vertical hoofs and concave soles, so admirably disposed to traverse rocky and slippery surfaces, is physically incompetent to exist in low-lying swamps. The unwieldy animal, slow paced and torpid, with a foot perfectly adapted to such a region (its ground face being so extensive and flat that it sinks but little, and the frog developed to such a degree as to resemble a ploughshare in form, which gives it a grip of the soft, slippery ground), is but indifferently suited for travelling on a hard, rugged surface.

'In process of time, however, *the small, concave hoof expands and flattens, and the large flat one gradually becomes concentrated, hardened, and hollow*, these changes being designed to suit the altered physical conditions in which the animals are placed.'—(Prize Essay, 1870.)

The above extract from Professor Fleming's essay shows that the conversion of a soft hoof into a hard one is simply a matter of time, and that the owner of a soft-hoofed horse must not be discouraged if the process should not be as rapid as he expected.

There is yet another point to be considered. Professor Fleming goes on the supposition that the two horses possessed untouched hoofs; and it is evident that if the hoofs had been mutilated by the farrier, the process must take a still longer time. Here again, owing to the variability of hoof-growth in different animals, it is impossible to lay down any definite law on the subject.

Taking the average of growth, the entire hoof of a shod horse is renewed annually. But when the hoof is worn away by friction upon hard ground, the growth is faster, in order to keep pace with the waste. Keeping these two facts in mind, remembering that all the nail-holes must have grown out, and not forgetting the idiosyncrasy of the animal, the trainer can form a very fair judgment of the time which must elapse before the hoofs can do their full work on hard roads. That the softest and worst hoof *can* be strengthened for shoeless work is shown by Mr. Baker's experience and challenge.

Let me again remind the reader that I am dealing with facts and not with theories.

Here are more facts.

The manager of one of our tram-car lines wrote, under the *nom de plume* of 'Humane,' a very remarkable letter, which is given in full in Appendix K of 'Horses and Roads.' I possess the real name of the writer. He begins by saying that he has to manage a working staff of thirty horses, whose work is on stone-paved roads, and who have to run about eighteen miles per diem, at an average rate of six miles an hour, stoppages included. Each of the shoes which these horses formerly wore weighed nearly two pounds, and the animals very soon took to brushing and cutting, and then began to 'go over on the knees.'

Suspecting that the fault lay in the shoe and not

in the horse, 'Humane' procured a set of Charlier shoes, and tried them on one of the horses. They were ten-inch shoes—*i.e.* more than double the needful length. The horse was run for nearly two hundred and thirty miles as a test, and the plan answered so well that the ordinary shoes were abandoned and the Charlier substituted.

'Humane,' however, was not content with the ordinary Charlier, but tried a series of experiments, the result being that 'the shorter the iron, the better it answers.' So he now limits the length of the shoe to four inches, its weight, nails included, being four ounces, as against the thirty-two ounces (without nails) of the shoes which were formerly used.

He makes his own shoes in the following manner: 'I buy the half-inch round iron and flatten it to three-eighths by half an inch. I cut off four inches, weighing four ounces, and nail on with No. 6 countersunk nails.'

Here I present the reader with an illustration copied from a photograph taken from one of 'Humane's' tram-car horses. The animal was one of the worst 'screws,' and was condemned to the knacker on account of his hopelessly diseased feet. His master, however, determined to give him a chance of life, and tried the Charlier plan. The horse at once went better for the change, and soon ran perfectly sound. When the ten-inch shoe (shown in the photograph) was removed, a four-inch shoe was substituted with the best effects. Lastly, when the shoe had to be taken off, 'Humane' tried the experiment of leaving the hoof entirely without protection. This battered, condemned, and hopelessly diseased 'screw' then ran five hundred and forty miles without shoes, and did his work with more ease than ever.

HOOF OF TRAM-CAR HORSE ('HUMANE').

Finding that his worst horse did so well, 'Humane' removed the shoes from three others, the result answering all his expectations. Unfortunately, horses, like men, have reasons to pray to be saved from their friends. Some well-meaning but ignorant persons raised the cry of cruelty, with all the usual nonsense about artificial roads compelling the use of artificial protection, and 'Humane' was obliged to have the horses again shod. He did not, however, return to the ordinary shoe, but employed the tip, which, though it may do no good, does less harm to the hoof than any other form of shoe.

I regret to say that I have heard of more than one case where the managers of the R. S. P. C. A. have openly declared that they will prosecute anyone who rides or drives a horse without shoes.

They refuse to make themselves acquainted with the structure of the hoof, to listen to arguments, or examine proofs. Mostly they have made up their minds that to drive a horse without shoes over hard roads would be as cruel as to make a man take off his shoes and run over the stones at once.

Sometimes they have consulted farriers, grooms, &c., and of course have been told that for a horse to traverse roads unshod would be impossible, because if the hard iron be worn away by friction, the comparatively soft horn could not possibly endure the work. Thinking that people who have been employed about horses all their lives could not be mistaken in such a subject, they accept the statement as if it were the result of experience based on facts, and were not, as it really is, a mere opinion of a subject in which they have had no practical experience.

## CHAPTER XI.

Hardening and renovating the hoof—Variety in hoofs—Thrush concealed by shoe—Dr. Brierley's horses—Horses in Italy—Mr. Theodore E. Williams's horse 'Prince'—Lame when shod—Experiment on another horse, and result—Mr. Herbert Smith's experiments—Altered shape of hoof—Need of perseverance—Xenophon's rules for hardening the hoof—General summary of the subject.

WE will close this portion of the work with a more detailed description of the process which a horse's hoof must undergo in order to enable it to do work upon our roads without any artificial 'protection.'

It seems rather curious that we should want to protect exactly that part of a horse's structure which Nature has triply protected. It is still more curious that the means which we employ for this purpose are such that it is impossible to use them without injuring and mutilating the very structures which we are trying to protect.

If all roads, and all hoofs, and all horses were alike, the process would be comparatively easy. But, as everyone knows, there is a vast variety in roads, and there is still more variety in hoofs and horses.

As to the hoofs, they vary in shape, in texture, in the rapidity of their growth, and in the angle at which they are set on the ground. There are round hoofs and long hoofs, high hoofs and flat hoofs, tough hoofs and brittle hoofs, &c.

Then, as it is very difficult to procure a horse whose hoofs have been untouched, it is necessary to take into consideration the amount of mutilation to which they have been subjected, and the consequent caution that will be required in the treatment.

Again, as has already been pointed out, even if no interior ailment be manifest, incipient thrush may exist, and exhibit no symptoms until the removal of the shoe and the pressure of the hoof upon the ground cause it to be detected. (N.B.—When this is the case, there are sure to be some who will say that the thrush was caused and not detected by the removal of the shoe.)

Thus, although the general structure of the hoof is alike in all horses, and the principle in which it is to be renovated is the same, the details must be subject to considerable variation, the amount and kind of which must be left to the discretion of the owner.

On referring to page 164, the reader will see that Mr. Whitmore Baker gives a minute description of the method which he employed while hardening the hoofs of 'Stella.' With the exception of the special 'preparation' Mr. Astley used similar means when dealing with 'Tommy's' hoofs; and, although three hoofs of his horse were unsound, succeeded, after several apparent failures, in rendering the animal able to do exceptionally good work upon hard roads without any artificial protection to the hoofs.

Dr. Brierley, of Manchester, who drove his unshod horse for years over the rough granite roads of that city, used no means except a gradually increasing exercise on hard ground. I have seen and handled the horse, which was then in full work, and all its hoofs were absolutely perfect.

Dr. Brierley has now a thirteen-hand pony, which, when purchased, was about four years old, had never been shod, but had not done much work. However, as Dr. Brierley writes in a letter to me (September 7, 1884), 'I ran him for a month without shoes, and he ran perfectly well, and *stepped*. He was absolutely sound, and the hoofs perfect. I contemplated selling him, and had him shod with Charliers, and ran him in that way up to a month ago, when he fell, barking both knees, while passing through Tarporley, in Cheshire.'

The pony was now again worked, and when the set of shoes had worn out, the hinder pair were not replaced. (I may here mention that horses in Rome and Naples, where the pavement is made of lava, which in wet weather is as slippery as ice, are scarcely ever shod on the hind feet, while many are not shod at all. In some parts, where the road is very bad, there are notices requiring all riders of shod horses to dismount.) But Dr. Brierley mentions that the space between the frog and the sole in his pony was so large that small stones became jammed in the angle, so that he was obliged to have it filled up with leather. Now, it is evident that the farrier's knife must have been at work, and that the Charliers were complete shoes and not tips, as they ought to have been. Had the hoof been let alone, there could have been no space to fill up, and had there been no shoe, there could have been no angle for the lodgment of stones.

It is but natural to ask why a horse which was in full work, which had all its feet sound, and which, in addition, possessed the much-coveted 'action,' should have been shod at all. The reason was, that the animal was intended for sale, and that at the present day purchasers can hardly be found for an unshod horse. They are so much accustomed to the shoe, that if an unshod horse were offered for sale, they would jump to the conclusion that there was something wrong about the hoof, and that the owner did not dare to put on a shoe.

It is quite certain that in the case of the Manchester pony, the animal's troubles began after the shoes were applied, and that for some of them, such as the lodgment of stones, the shoe was the sole cause.

I will give two more examples of horses that work without shoes.

Having heard that Mr. Theodore E. Williams, of Salterley Grange, near Cheltenham, had for some time freed his horses from shoes, I wrote for further information, and received the following answer :—

'September 15, 1884.

'In reply to your request for information respecting my unshod horses, I may observe that I first discarded shoes about three years ago in consequence of a hunter called "Prince" being slightly lame a day or two after hunting in the dry weather of March, which I attributed to the concussion with the iron shoe on hard ground.

'Considering how this was to be avoided, I felt that the iron was harder even than the ground over which the horse had to travel; I therefore determined to remove the shoes, and to allow the horse's hoofs to recover their natural shape and condition, in which, of course, there would be no nail-holes. So I placed the horse in a straw yard, and fed him with vetches and clover. As his hoofs grew, they were pared and rasped, until the nail-holes had entirely disappeared.

'He was then transferred to the stable, and exercised as usual, but I soon found that his feet wore away faster than they grew. In trying to restore the horse to its natural condition, I was asking him to travel over roads almost as hard as rock, while for twenty hours out of twenty-four he stood on litter as soft as grass. It was like a hand accustomed to be gloved working the ropes of a ship, or an English gentleman's child trying to race barefooted on the granite roads with the shoeless children of the Scotch Highlands.

'I therefore *had the litter removed by day*, and henceforth the horse's feet became as tough and hard as oak, and he has ever since travelled with ease and elasticity. My companions in the hunting field are often astonished to see how readily he gallops along a road, even when fresh stoned; and when, as occasionally happens, he has to jump a fence into a road, he does so with much less " jar " to his legs. In difficult places I am sure he is more active and quicker in recovering himself than when shod, and *he has never been lame since his shoes were removed.*

'Many of my friends said, "Ah! you have certainly succeeded with *that* horse, but then he is light, and must have exceptionally good feet. You could not do the same with a heavy horse, as he would knock his hoofs all to pieces."

'I therefore, the summer before last, removed the shoes from a sixteen-hand six-year-old horse, well up to sixteen stone, and turned him out. I frequently rasped his feet until the nail-holes disappeared, allowing the wall of the hoof to project very little beyond the sole.

'He ran throughout the summer, but was often otherwise worked, although his feet were not specially hardened for it. For instance, I give three days' work in succession. First day, sixteen miles, ridden fast on the road. Second day, driven seven miles in single harness, drawing five or six people. Third day, ridden twenty-four miles on the road. Throughout the summer I used him whenever I wanted an extra horse. Last winter, I hunted him in his turn with as satisfactory results as those before mentioned with regard to "Prince," and this summer he has been kept in condition, and regularly ridden and driven.

'In addition to the horse's action being more elastic, and therefore much pleasanter to ride, I consider he is safer, and it is obvious that I avoid the risk of nail-pricks, corns, over-reaches, seedy toes, contracted feet, and probably of splints and side bones; I therefore think that these horses are likely to work much longer than if shod.'

This letter is a peculiarly valuable one, because it contains a practical answer to almost every objection that can be made. That the horse ' Prince ' was an exceptional animal is answered by the success with the second horse. That an unshod hoof would be cut to pieces on a newly-mended road is disproved by practice; and, as the same horse was used for hunting across country, for riding over roads, and for driving a carriage in single harness, the unshod hoof is shown to be equal to any kind of work.

One more instance of the power of common sense and perseverance.

The following letter, narrating the history of a horse transformed from a shod into an unshod animal, is extremely valuable, as it shows what can be achieved by common sense when matched against prejudice :—

'Marton, Rugby: June 5, 1882.

'Dear Sir,—In accordance with my promise, I write you the result of driving my horse without shoes.

'He was rising five years, and had been shod ever since he was handled. He was a bad subject for the experiment, as he could never bear the frog to touch the ground. I have now driven him for six months, and he travels much easier to the driver, bears thoroughly on the frog of his foot, and is more sure-footed. I do not intend him to be shod again, and had I a colt, I should never shoe him. The chief objection to removing the shoes from a horse accustomed to them is the expense of keeping him idle until the hoof begins to assume its natural shape.

'I was fortunate in having a groom who endeavoured to carry out my ideas in spite of his own prejudices.

'Any person trying the experiment should insist upon thorough cleanliness in his stable, look after the horse himself, be endowed with perseverance, and *armed in proof against ridicule*.

'I may say I drive without bearing-rein or blinkers. This latter plan requires great care until your horse is accustomed to it.

'I will not intrude longer upon your time, but should any person wish to question me, I shall be very pleased to reply to them.

'Yours truly,

'Herbert Smith.'

Taking Mr. Smith at his word, the late Mr. A. F. Astley wrote to ask him if he would let him have some details concerning the mode which he adopted in fitting the hoofs of his horse for road work. He was kind enough to send at once the following answer, dated June 9, 1882:—

'My horse had worn a set of shoes for about three weeks when I had them taken off without preparing his foot in any way, and he was left unshod at once (no tips).

'For the first four or five days he stood in his horse box, only littered down at night; all litter was taken away during the daytime, and all refuse cleared up as soon as made, so as to prevent his hoof being softened by standing in it.

'For the following week or ten days he was loose in a paved stable yard, to stand or move about as he chose. He was then led out on the turnpike road (not the turf) daily, at first for only three or four hundred yards, the distance being gradually increased. As he travelled better, he was mounted or driven in a trap, but for some time restricted to a walk.

'In about three months I began to trot him for short distances, and thus he was gradually able to undertake his ordinary work. I do not think his action is so high as when shod, but he is more surefooted, pleasanter to drive, and has a healthier hoof than before. He appears not to travel so fast as he did, but this is only in appearance, for on timing him I find he is fully as quick on his journeys.

'His hoofs assumed a very ridiculous shape while attaining their present form, the coronet at one time projecting like a moulding above the lower crust (or wall). The crust of the hoof has only cracked where the old nail-holes had injured it, and as they wear away it becomes sound and hard. (See page 107.)

'The whole of the tenderness of the foot had proceeded from pressure on the frog. Only once was there a suspicion of its being worn tender, and if it really were so on that occasion, a single day's rest in the stable cured it. Of course his feet were well washed, and I examined them myself each day on his return from exercise.'

In this exceedingly valuable account, the attention of the reader must be drawn to one or two points.

One of them is, that the owner of the horse ought not to be discouraged even by such an unexpected phenomenon as an altered shape of the hoof. Another, that although of course it would be better for a horse never to have been shod, provided that its hoofs were rightly treated, the fact that it had been accustomed to wear shoes does not prohibit it from doing without them. A third point has to be noticed—namely, the loss of the animal's service for some months; and a fourth is the extremely judicious character of the means which were employed.

Let us take the very important subject of the loss of services, so that the horse is 'eating his head off' while his hoofs are hardening. Few owners of horses can afford to keep the animal idle, and there is not the least necessity for it.

The simplest plan is to wait until the time comes for shoeing the horse, and to substitute a Charlier half-shoe for that which was removed. Then, as has already been mentioned, if the size of the Charlier be reduced each time of shoeing, the horn will have become so strong and hard, that the small tip to which the shoe will be reduced about the third or fourth time of shoeing may be removed altogether, and nothing substituted for it.

Next we come to the means which were employed in this particular case. Nothing could have been more judicious, and any one who follows the same plan will produce the same result.

The plan, however, is not a new one. It was practised and recommended nearly 2,300 years ago by Xenophon, who, as the reader will remember, was not only a historian, but an acknowledged authority on horses and dogs, and the general who conducted the celebrated retreat of the Ten Thousand. In so doing he was obliged to organise cavalry as he could, taking any horses on which he could lay his hands. That most trying march lay over all kinds of soils, and in order to enable the hoofs of the horses on which so much depended to endure such an ordeal, he laid down the following rules:—

'To prevent stable floors from being smooth, they should have stones similar to a horse's hoofs in size inserted in the ground, for such stable floors give firmness to the feet of horses that stand on them.

'The groom must also lead the horse out of the stable to the place where he is to comb him; and he should be tied away from the manger after his morning's feed, that he may come to his evening's meal with the greater appetite.

'The ground outside the stable may be put into excellent condition, and serve to strengthen the horse's feet, if a man lays down here and there four or five loads of round stones, each large enough to fill the two hands, and weighing about a pound, surrounding them with an iron rim, so that they may not be scattered. For, as the horse stands on these, he will be in much the same condition *as if he were to travel part of every day on a stony road.*

'A horse must also move his hoofs while he is being rubbed down, or when he is annoyed with flies, as much as when he is walking, and the stones which are thus spread about strengthen the frog of the feet.'—(Xenophon's 'Hipparchicus,' *i.e.* Horsemanship, ch. iv. par. 3, 4, 5. Watson's translation.)

Let us now try to act as a judge acts in summing up a case at law when there is conflicting evidence. He distinguishes between facts and surmises, refusing to allow the latter to appear as evidence. In the case of actual evidence differing, he impartially considers both sides, and delivers his judgment according to the value and weight of evidence.

The first and chief argument in favour of the iron shoe is, that the horse was intended for soft soil, and not to wear out his hoofs by carrying heavy weights or by drawing heavy carriages. Also, that as we use hard and artificial roads, the horse's hoofs need artificial protection.

*Per contra*, it is urged that these statements are mere assumptions, both of which can be contradicted.

In the first place, the original home of the horse is in Central Asia, where the soil is hard and rocky. In the next place, the worst of our artificial roads is far easier for the hoof than the broken ground which the wild horse traverses. In the next place, it has been proved, and is an existing fact, that unshod horses can do regular and hard work whether they carry a heavy rider or whether they draw a heavy carriage, and that they can work upon the roughest, hardest, and most trying roads in England.

It is said that because a horse goes lame if he casts a shoe, the necessity for shoes is proved.

Not at all. If the hoof were left uninjured by the shoe and its adjuncts, such a theory might be tenable. But if, in order to put on the shoe, the farrier weakens the hoof with his knife, rasp, and nails, the argument falls to the ground.

It is said that some horses which were worked unshod went lame.

Very likely they did, but not if the hoof had been allowed to harden to its natural consistency. Mr. Astley, for example, failed at first with his horse 'Tommy,' and was on the point of giving up the experiment in despair. But the fault lay not with the horse or the hoof, but with the owner, who was too impatient, and did not give the artificially weakened hoofs sufficient time to harden. As the reader may remember, the experiment did finally succeed, although the hoofs were in a very enfeebled state when the shoes were first removed.

Do shod horses never go lame? And when they do, it can scarcely be accepted as a proof that every horse must have its shoes pulled off. Neither, when a barefooted horse goes lame, is it a proof that every horse ought to be shod. Moreover, there are many living examples of horses which are always lame when shod, and always go sound as soon as the shoes are removed.

In a letter dated January 14, 1882, Mr. A. F. Astley sent me a piece of information which is worth recording :—

' A 'bus driver with whom I went yesterday made some striking statements incident to the shoe question.

' He was still a young man, and had been in a cavalry regiment, where he had the charge of a very vicious mare. Shoes seemed to be her aversion, and she would kick and kick until she got them off or loosened them. She once did this on the march, but though with her feet unprepared for road work, she completed her march without shoes.'

Mr. Astley also sent me a few lines which effectually dispose of the theory that the horse was intended to live on soft ground :—

' Yesterday I was taught much ! A roan, never shod, three years old, having been advertised in the " Bazaar," I went down to see him.

' There could not be a more striking proof that *hard ground is wanted to keep the hoof in form.* This colt had been kept in a grass field — a dry one—and his hoofs had grown very long and out of all form. In this weakened state he had got out of the field and galloped on the road. So the hoofs are cracked, and the cracks have extended.'

The reader will perceive how an advocate of the iron shoe would seize on this fact as a proof that horses must be shod in order to enable them to carry on their work. Whereas, if the horse had been in the hands of anyone who understood the anatomy of the hoof, and had been taken over a few miles of hard roads daily, it might have traversed all the roads in England, from the flats of Cambridge and Essex to the hills of Devon, Stafford, and Derbyshire, and have possessed better hoofs than when it started.

My vocation as lecturer takes me over all kinds of roads in all parts of England and Scotland, not to mention America, though I cannot say, as Dickens did, that I have been upset out of every imaginable conveyance that goes on wheels. Still, I do happen to possess practical knowledge of the roads of Devonshire, Derbyshire, and Staffordshire, all of which are traversed daily by unshod horses doing any kind of work. Here is an account of the ordinary roads of North Staffordshire, extracted from the 'Sporting and Dramatic News,' an acknowledgedly 'horsey' authority. It is dated August 9, 1884, and signed by 'Rapier,' whose identity is well known to all sporting men :—

'It is reserved for Staffordshire, or at all events North Staffordshire, to be distinguished amongst all other counties for having quite the worst roads in England. I do not think I am far out in saying that they never bind at any season of the year.

'Possibly, the only " metal " that is to be obtained locally consists of the round pebbles (if a stone which is sometimes nearly a foot in diameter can be called a pebble) which are from time to time scattered on the roads. But I should have thought that somebody might be found to break them into pieces. Being round and smooth, they have, of course, no sooner worked in with the wear of the traffic than they wear out again.

'The moral of this is that people who live in Staffordshire should not drive delicately built carriages. I can recommend, however, carriage exercise in this neighbourhood for those who are so singularly constituted as to have livers.'

North Staffordshire, in fact, is very much in this country what New Hampshire (the 'Granite State') is in America, as I know from practical experience in both localities. Yet an unshod horse has been driven over these roads without suffering any injury to the hoof.

There is, then, only one merit which can be urged in favour of the iron shoe, namely, that of protecting the hoof, and, as we have seen, there is abundant evidence that the hoof needs no such protection.

Now we will take the opposite side, and see what can be urged against the iron shoe nailed on the hoof.

It is not only asserted and surmised, but proved, that the shoe is the direct, or indirect cause of every disease to which the hoof is liable. It causes Corns, which could no more exist in the hoof of a shoeless horse than in the foot of an Australian savage who never saw a shoe in his life. The shoe causes laminitis, quitters, thrush, and navicular disease, all being inflammatory in their nature. Contracted hoof, greasy heels, and sand-crack are equally attributable to the shoe, and make the very name of farrier a terror to all who care for the welfare of their horses.

Then, the unpleasant habits of cutting, brushing, interfering, and clicking cannot be contracted by horses which do not wear shoes. The profession is so well aware that these faults are due to the shoe, that a whole class of shoes has been constructed, each of which was supposed to have the merit of obviating one or other of these defects.

I do not make any of these statements on my own very feeble authority. Every one of them is to be found in such writers as Fleming, Mayhew, Lupton,

Youatt, Douglas, Miles, Bracy Clarke, Fearnley, Ransom, Treacy, Kendall, &c. Some of these writers are English and some American, and their works are easy of access in any good public library.

Then, the perils of the nail have to be considered, 'pricking' the foot always causing great pain and lameness, and sometimes ending in the death of the animal.

The pecuniary saving of abandoning the shoe I do not take into consideration, neither the cost of veterinary attendance on horses which have been injured by the farrier. The saving is of course considerable, but my dealings are with the welfare of the horse, and not with the pockets of the owner.

In concluding this portion of the work, I only again ask the reader to balance the weight of actual evidence on both sides of the question, to discharge from his mind all prejudices, and to deliver his judgment as impartially as if he were a judge acting in a court of law.

## CHAPTER XII.

The Professional Eye—Fashion and nature—The curb—Weight and size of bit—The BEARING-REIN—Three kinds of bearing-rein—The gag bearing-rein—Mechanical parallel—The over-head rein—Neck of the horse—Great ligament of the neck and its attachments—Vertebræ of neck and spine—Vertebræ and railway buffers—Arrangement of a train—The martingale—Rattling of harness and tossing of heads—Sir Arthur Helps' opinion—Effect of the gag bearing-rein on the spine and feet—The 'burr' bit of America—Mr. Henry Bergh's work—The locomotive and the horse.

WHAT a wonderful product of civilisation is the Professional Eye! It begins to develop itself as soon as man emerges from pure savagery, contenting itself at first with the nose bone of the Australian, the lip-disc of the Botocudo, and the tattooing of the Marquesan and New Zealand chief. It rests with satisfaction upon the vagaries of fashion, upon the furnishing of our houses, and the decoration of our gardens. Therefore, that our horses should be subject to its sway is only to be expected.

A few years ago, I met with a treatise on the management of horses, in which the writer gives it as his opinion that a thoroughbred horse, properly harnessed, is the finest sight in the world. This statement is illustrated with a figure of a 'properly harnessed' horse. To my very unprofessional eye the horse would have been a much finer sight without the harness, which, in fact, becomes in the illustration the primary object, the horse enclosed within it being of secondary consideration.

There are blinkers to prevent the horse from using its eyes. There is a 'gag' bearing-rein to prevent the animal from putting its head down, and a martingale to prevent it from throwing its head back, and there is a long-cheeked curb bit, together with its corresponding chain. Every one of these appliances is not only useless, but actually injurious to the horse, lessening its powers of work, and wearing it out long before its time. By way of a contrast, I here give two figures, copied (with a few slight additions), by permission from the late Mr. E. Fordham Flower's valuable pamphlet entitled 'Bits and Bearing-reins.' The reader would hardly imagine—I did not do so until told by Mr. Flower himself—that the figures are facsimiles of the photographs of the same horse! One represents the animal as it appeared when purchased by Mr. Flower, and hampered by improper harness, and the other represents it as it appeared when driven by Mr. Flower a few weeks after purchase.

Of this animal Mr. Flower writes as follows:—

'A few years ago I bought a fine horse with a bad character—he was a rearer, a jibber, a bolter—and the late coachman told me I should never be able to drive him. But I liked his looks, and the

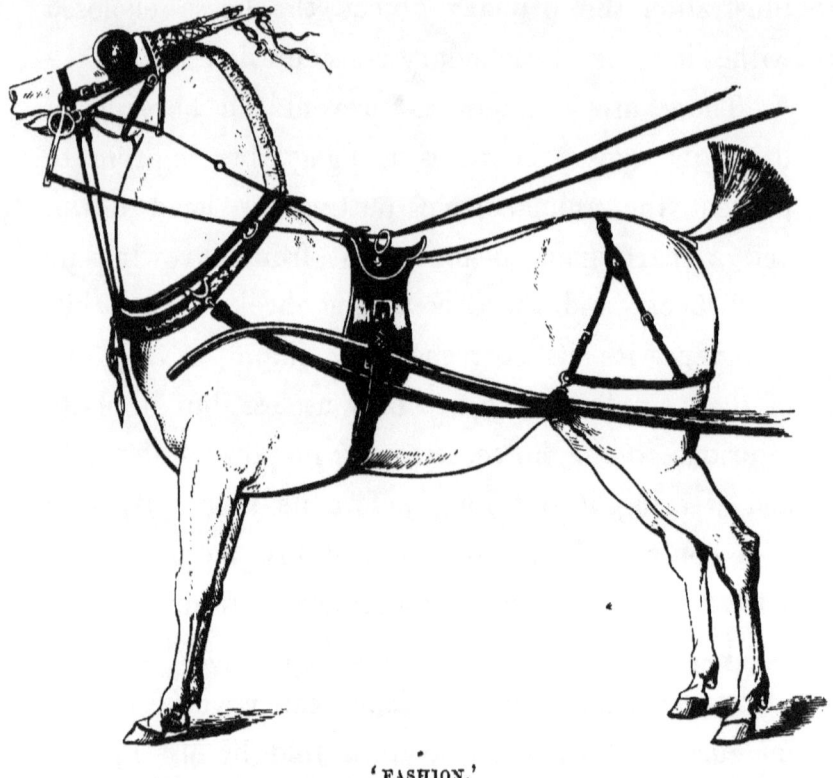

'FASHION.'

[The attitude and harness of this figure are copied from a photograph of Mr. E. F. Flower's horse as it appeared when he bought it. The ears, mane, and tail are explained in the text.]

result confirmed my good opinion.' How Mr. Flower wrought this change we shall now see.

We will first look at the bit.

## MR. FLOWER'S HORSE.  201

The amount of iron which is too often used in a bit is almost incredible. On the next page is a figure of the bit which was worn by Mr. Flower's horse when he purchased it, and which was pro-

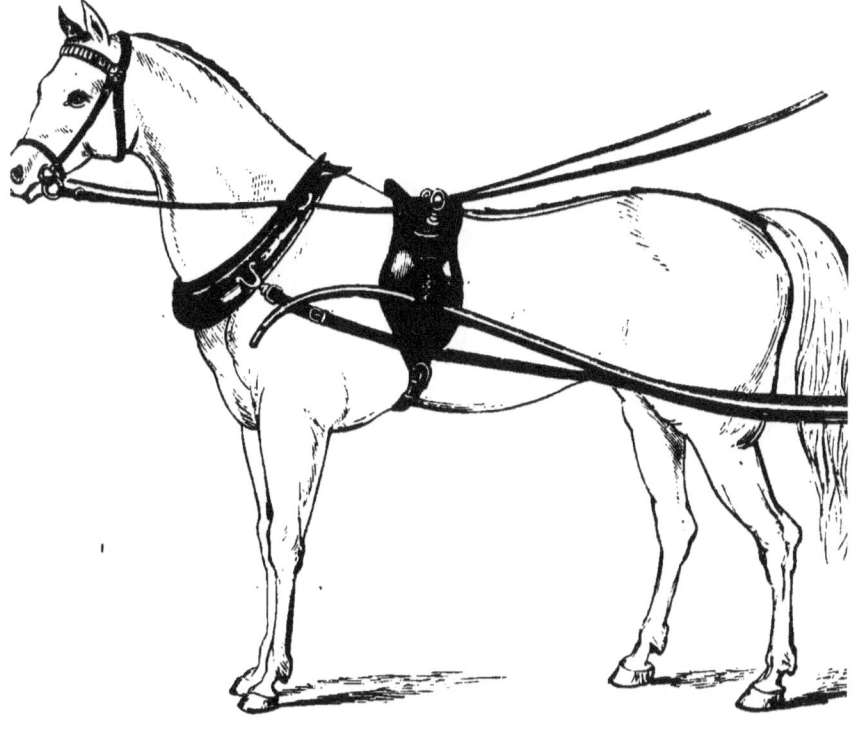

'NATURE.'

[This is from a photograph of the horse 'Fashion' as it appeared after Mr. Flower had possessed it for a few weeks.]

nounced necessary in order to subdue its bad temper, and prevent it from running away. The machine weighed two pounds all but two ounces, and it seems almost incredible that such a piece of ironwork should be forced into the delicate mouth of the horse, but I have seen and handled it repeatedly.

Such a machine may, indeed, coerce the horse to a certain extent, just as is the case with the terrible Spanish bits, which are powerful enough to smash the jaw of the animal. But the horse was never meant to be coerced, but to be the willing servant of man; and there is, in every high-spirited horse, a point at which coercion loses its power, and the horse, despite the pain inflicted upon it, becomes a rebel. How often do we not hear the warning given, 'He goes quietly on the snaffle, but if you touch the curb, look out for squalls.' The moral of which ought to be, that the curb should not be used at all.

THE BIT WHICH WAS REMOVED BY MR. FLOWER.

A year or two ago I was accompanying a lady in her carriage, which, as the day was wet, was a closed one. The movement of the carriage was anything but smooth, and at last became so jerky that the lady asked me what could be the matter. Being thus appealed to, I said that I was sure that the off horse was curbed up too tightly, and that the irregularity would continue until the animal was released from the curb.

She immediately stopped the carriage and went to the horses' heads. The near horse was all right, but the curb-chain of the off horse was drawn so tightly that it actually sank into the skin of the lower jaw. It was so tight that she could hardly unloose it. On her inquiry, the coachman—a new one—said that the horse was so restive, and given to bolting, that he was afraid to drive it except with a tight curb.

However, to his amazement, the lady not only loosened the curb, but took it off, and put it in her pocket, so that it could not be replaced. Then she shifted the reins to the snaffle rings, and returned to the carriage. The horse shook his head three or four times, as if to assure himself that he was relieved from the curb, and then went off with perfect ease and smoothness.

I had a talk with the coachman afterwards, and found, as I had surmised, that the man had only followed the traditions of his order, and thought that a high-spirited horse must be a vicious one. He was entirely surprised at the result of his mistress's action, and when she took her place in the carriage after removing the curb, he made up his mind to be prepared for an accident.

I am sure that the lady in question will pardon me for recording this historiette, because it redounds very much to her credit, and I mention no names.

I feel quite certain that we do not realise the horrible pain which is given to horses by these enormous and weighty bits.

The reader will remember that the two portraits on pp. 200 and 201 were taken from the same horse. That on p. 201 was taken first, so as to show his appearance when driven with a simple snaffle. After this portrait was taken, the harness was removed, so that the horse might be shown as he was when Mr. Flower bought him.

No sooner did the animal see the groom approach with the old bit and curb than he began to tremble, and burst out into a profuse perspiration from absolute terror. The effect upon the horse was so great that Mr. Flower could scarcely bring himself to order the bit and harness to be placed upon him.

The reader may perhaps think that this bit is an exceptional one, and made especially for that horse. I only wish it were so. As I write these words, September 1884, there is in the Crystal Palace an International Exhibition of Arts and Manufactures, and among the various stalls are several which are devoted to harness, and in which may be seen bits which are even more ponderous than that which caused such torture to Mr. Flower's horse.

There is another point to be considered with regard to these heavy bits, and that is, the pain which the cold iron causes to the animal in the winter time when the bit is first put into its mouth. Even in ordinary weather the contact of cold iron with the teeth will cause pain, but in the winter time it must be absolute torture.

All these bits and curbs are visible proofs that as a rule those who ride or drive, and especially the latter, seem to think that the object of the bit is to coerce the animal, whereas its right object is to guide it. A horse which has been trained to act as a servant and not an unwilling slave, does not need to have his jaws dragged this way and that. All that he requires is that his master should let him know what is wanted of him, and he will be only too glad to obey. We shall revert to this branch of the subject on a future page.

### THE BEARING-REIN.

CRUEL as may be the bits which I have mentioned, they are but trifles until they are combined with the Bearing-rein, an invention by means of which the lightest snaffle may be transformed into an instrument of torture. In America it receives the more appropriate title of the 'check rein.'

Originally, there is no doubt that the bearing-rein was intended for purposes of show, so as to hold up the horse's head, and give him an appearance of being a horse of spirit.

The custom is of very great antiquity, as is shown by the ancient Egyptian monuments. There we see Pharaoh represented as standing in his chariot of state, guiding his armies to conquest, and receiving the homage and tribute of the vanquished. The horses are fitted with driving-reins and bearing-reins, the latter being hooked to the saddle exactly like ours of the present day.

The sculptor has hit off with curious fidelity the peculiar action of a horse thus trammelled, and marked strongly the distinction between the harnessed horses of state and those which are ridden by the combatants and have their necks free.

Two kinds of bearing-reins are employed in this country, and a third is largely used in America.

First we have the ordinary bearing-reins, which are fastened to the cheeks of the bit, then pass through a couple of drop-rings, one on each side of the head, and are finally hitched over a hook or peg which is fastened to the saddle. There is just one advantage which this rein possesses—namely, that when the horse is at rest, or when it is ascending a steep hill, the driver can unhitch it.

The other invention is sometimes called the 'Bedouin' and sometimes the 'Gag.' The latter is far more appropriate than the former, as no Bedouin, even if he uses a cruelly powerful bit, and will wrench the animal's jaw severely in executing the various feats of horsemanship in which those Arabs take a pride, ever invented such a piece of machinery as the Gag bearing-rein.

The accompanying illustration is copied by permission from the work to which reference has been made. The original was taken from a photograph.

In the first place, this rein is quite independent of the driving bit, and has a bit of its own. The rein is first attached to the head-stall, as seen at D. Then it goes through a swivel attached to its own bit, as is shown at C. Thence it passes through the drop-ring at B, and then is hitched over the hook attached to the saddle at A.

Now let us see what is the effect on the horse.

The small diagram which is appended to the figure shows this rein as reduced to its mechanical equivalent, and has the same letters. When a rope, weights, and pulleys are arranged as seen in the diagram, a force which is represented by one pound at A will exactly balance a weight of two pounds at C, so that its power is doubled. Now, in the Gag bearing-rein the power is represented by the hook on the saddle, which, by the way, is fastened to the tail by a crupper. B is the fixed, and C the movable

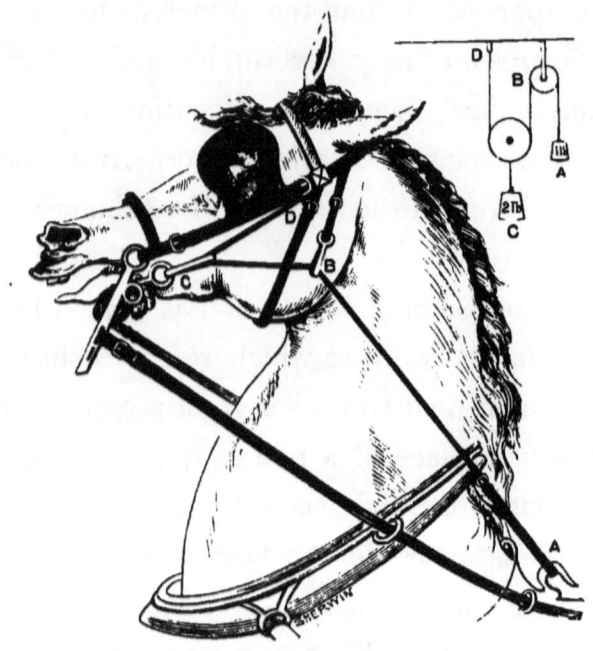

'BEDOUIN,' OR 'GAG,' BEARING-REIN.
(From Mr. E. F. Flower's book.)

pulley, while D is the fixed end of the rope. Consequently the force of the rein at A is exactly doubled at the horse's mouth or C.

Unlike the simple bearing-rein, this machine cannot be unhitched in order to release the tension of the neck, as the bit is quite independent of the driving bit, and would fall out of the horse's mouth when the rein was unhitched.

How this bit is applied may be seen from the following extract from Sidney's 'Book of the Horse:'

'Your London coachman of the highest fashion begins by drawing up the gag bit until he has enlarged the mouth by at least a couple of inches. He then adds a curb-bit of an inch too wide and four inches too long, quite regardless of the size of the horse's mouth, and having curbed this up tight, climbs on his box and makes, whether moving or standing at a door, a display very satisfactory to the distinguished owners, who have not the least idea that their horses are enduring agonies for hours.

'The result is shown by degrees in foaming, bleeding mouths, lolling tongues, spasms, restiveness, &c.' In other words, the horses are the victims of the professional eye.

The third kind of bearing-rein I have often seen in America, but never in England, and hope never to do so. Instead of the rein passing along each side of the head through a drop-ring, it passes over the top of the head, and then directly to the saddle hook. For this reason it is called the 'over-head check-rein.'

I have already said that the primary object of the bearing-rein was to give the horse a more spirited appearance. But two other reasons are invariably

P

brought forward where objections are made to its use. One is, that it keeps the horse on its feet and saves it from falling and breaking its knees; and the second is, that it prevents the horse from running away. We will take these three 'reasons' separately, but will first describe some more of the mechanism of the neck and backbone.

As the reader may remember, all the parts of an animal are mutually dependent on each other, and any interference with one of them will exercise an injurious influence on all the others.

It has already been mentioned that elasticity is one of the leading characteristics on which the structure of the horse is framed. How wonderfully elastic is the hoof in its natural state, and how the horse is injured by destroying that elasticity, has been already shown. Now it is evident that an elastic hoof would be useless if the body were rigid, and therefore, as might be inferred from the hoof, the neck and backbone combine elasticity and strength in a most wonderful manner.

This is not intended to be an anatomical work, and therefore I only give those details which bear immediately on the subject.

If the reader will refer to the illustration which represents the seven vertebræ of the neck, he will see that they are furnished with various projections.

Some of these are intended for the attachments of the wonderful piece of mechanism which is here shown.

We all know that at the back of the human neck

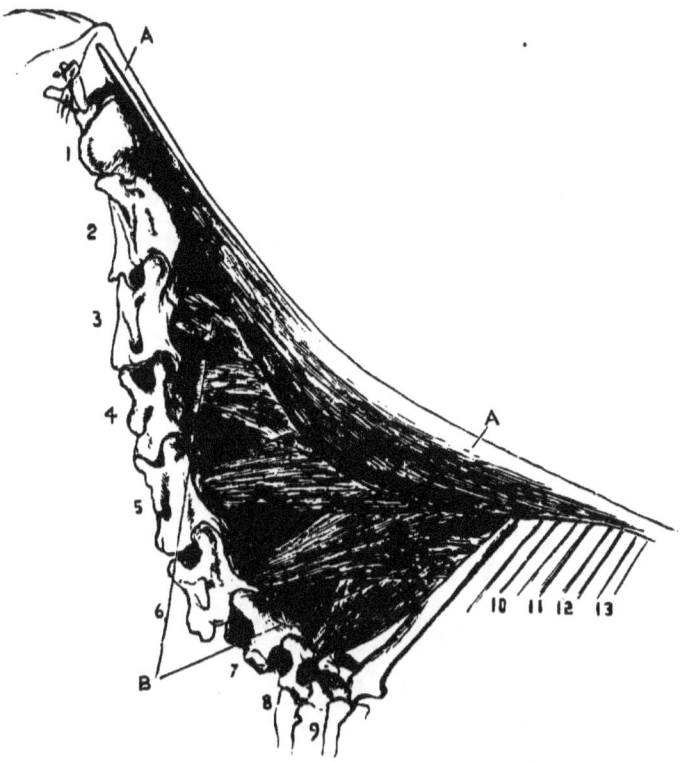

LIGAMENT OF THE NECK AND ITS BRANCHES.

there is a strong ligament called scientifically 'ligamentum nuchæ.' In the language of the butcher it is termed the pax-wax. With man, the weight of whose head is perpendicular, the ligament is, though strong, comparatively small and simple, and not intended to sustain the head in a horizontal position. This we can realise if we have dropped some small object and have been hunting for it on our hands and knees.

The horse, however, needs a much more elaborate ligament. At A is the upper and rounded portion which is attached to the head. Widening and thickening as it passes away from the head, it reaches the upper processes of several vertebræ of the back. In addition it sends out a series of branches which are attached to the vertebræ of the neck, as shown at B.

Still carrying on the train of reasoning, an elastic hoof and neck would be useless if the remaining

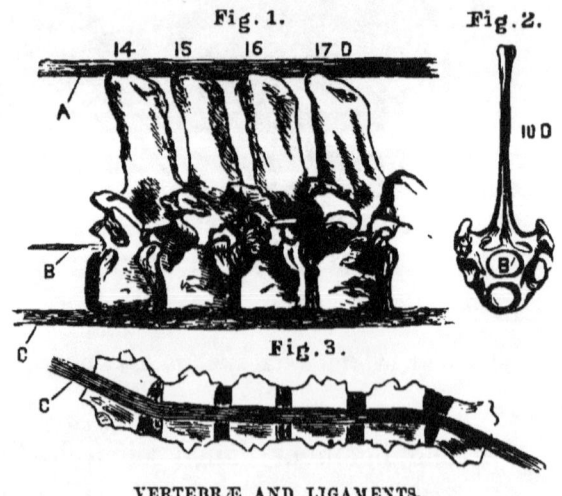

VERTEBRÆ AND LIGAMENTS.

portion of the spine were rigid, and so we find that the elasticity of the neck is carried through the rest of the spine. In the illustration we have seen how the ligament of the neck is attached to the vertebræ of the back. The preceding illustration takes the four vertebræ next in order. Fig. 1 shows how the vertebræ are connected by ligaments, one set running above and the other below, as seen at A and C; fig. 2 gives the front view of the tenth vertebra, and shows at B the hole through which the spinal cord passes, and on the floor of which rests another ligament. Fig. 3 is a diagrammatic representation of the vertebræ curved in action.

There is yet another provision to ensure the requisite elasticity. At each end of the 'body' of the vertebra there is a thick, rounded, and very elastic pad of cartilage, as may be seen diagrammatically shown at fig. 2 of the illustration on page 214. At A, the vertebræ are shown in their natural position, the elastic pads being in contact, and held together by the ligaments which have been described. At B,B they are represented as compressed to the utmost, so that their elasticity is practically annulled.

At fig. 1 is given a diagrammatic sketch showing the analogy between the chain of vertebræ and the present mode of coupling railway carriages.

When I used to go to school, in the days when railways were almost in their infancy, second-class carriages having only two ends and a roof but no sides, and the third-class carriages being mere wooden trays without seats, the system of coupling was in an equally primitive state. There were no elastic buffers, the only substitute being square wooden blocks.

The consequence was, that when the train started, a continuous series of jerks took place, each carriage jerking forward the one immediately behind it.

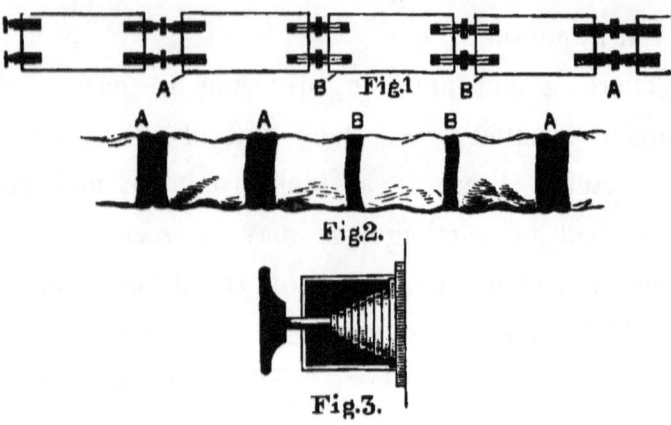

VERTEBRÆ AND RAILWAY BUFFERS.

When the engine slackened speed before stopping, a corresponding series of bangs ran through the train, each carriage being banged several times before it stopped.

This arrangement is now confined to goods trains, and is not likely to last, as even with them the wear and tear caused by this jerking and banging are very expensive.

Nowadays, instead of the solid wooden blocks, there is at each corner of the carriage a spring buffer, one of which is shown in a section at fig. 3. When the carriages are coupled, they are drawn together by double screws, so that the buffers are not only in contact, but press slightly upon the springs, as is seen at A. Thus, the train becomes a single body, instead of being a row of separate carriages, and a good driver can start or stop the train so quietly, that the passengers, if not guided by the sense of sight, can scarcely tell when they have started or stopped.

It is evident that if any of the carriages were to be screwed so tightly together that the buffers were forced back to their farthest limits, as shown at B, the elasticity would be destroyed, and the buffers rendered useless.

Now, a bearing-rein of any kind interferes with the wonderful system of elastic structures which have been described. Even if it be a very mild one, it hinders the play of the vertebræ upon each other. The horse's head being tied to the saddle by the bearing-rein, and the saddle being fastened to the tail by the crupper, the horse's head is practically tied to his tail, thus interfering with the elasticity which demands freedom throughout the entire length.

The gag-rein, however, does more, and forces the vertebræ together, like the overscrewed buffers. If the reader will refer to the figure entitled 'Fashion,' on p. 200, he will see that the whole attitude of the animal is quite different from that of the horse which is entitled 'Nature.' The peculiar position of the legs requires notice. It is not accidental, neither is it peculiar to the horse, the animal being the same individual as 'Nature,' though under different circumstances.

The fore-legs are pushed out in front, and the hind legs extended backwards, while the middle of the spine, below the saddle, is bent downwards. The reason for the attitude is this:—The head has been drawn back so far by the bearing-rein, that the pressure on the mouth is continuous, and its force doubled by reason of the mechanical action of the rein. In order, then, to relieve the mouth as much as possible, the horse instinctively stretches out its legs, and compresses the pads between the vertebræ, so as to shorten the spine, and thus to lessen the pressure on the mouth.

In most cases, although the head cannot be bent downwards, it can be lifted upwards, but the coachman who drove this particular animal (which afterwards went quietly in a snaffle, as in 'Nature') was so afraid of it, that he fastened its head down with a martingale, so that the head could neither be moved up nor down.

This, of course, was an extreme case, and therefore was selected by Mr. Flower as a 'shocking example.'

But even in those instances where the martingale is not used, the torture is extreme, as is shown by the action of the horses. Coachmen and grooms, who are essentially possessors of the professional eye, have a rooted idea that it is very grand to have their horses tossing their heads and flinging foam about.

They really think that the head-tossing and foam-flinging are marks of pride and spirit, whereas they denote pain, and vain attempts at alleviation. The horse finds that although he cannot stoop his head forwards without a severe jerk to his mouth, he can for a moment relax the pressure by throwing his head upwards. As to the foam, it is caused by the pressure of the severe bit. No one ever saw a horse, however spirited, or however high bred it may be, toss its head and fling foam except when it is tortured by the bit and bearing-rein.

Even if he were blind, a person who takes an interest in this subject can tell by his ears alone whether horses are wearing a severe bearing-rein. The peculiar rattling of the head-harness, which is thought so grand and noble, is simply produced by the attempts of the horse to shake the bit away from the tortured corners of his mouth.

It is pitiful in the 'season' to go through the Park, or to pass through the fashionable streets and squares, and to see the sufferings which are endured by horses. While being driven round the Park, stopping at fashionable shops, exhibitions, at the doors of their owners' acquaintances, or even at the door of the Royal Society for the Prevention of Cruelty to Animals, the horses may be seen undergoing this horrible torture for hours together.

The late Sir Arthur Helps put the point very forcibly, but not more so than it deserved. He lays the fault on the owner, who must be either ' utterly unobservant of what he ought to know, or pompous, or cruel.' He finishes as follows :—' I observe the equipages when the irrational tight bearing-rein is used. I then look on the arms on the carriage, and I know who are the greatest fools in London in the upper classes. The bewigged brute and idiot of a coachman of course thinks it a very fine thing to sit behind these poor animals with their stuck-up heads; *but his master ought to know better.*

'So ought his mistress.'

I have often wondered to see ladies sitting in their luxurious carriages, evidently ignorant of the fact that the incessant tossing of head and rattling of harness are sure proofs that their horses are suffering from ceaseless pain. A dog would yelp and a cat would scream if such pain were inflicted upon it, and so would force its tortures upon the ear when the eye took no notice. But the horse neither yelps nor screams. It suffers silently, and its owner is too thoughtless to see that it is in pain.

As to the white foam which is thought to be the result of exuberant spirit, the blood-stains with which it is often flecked ought to tell their own story to an observant eye, even though the spectator knew nothing of the habits of the horse. I regret to say, however, that the generality of carriage-owners know nothing of their horses, but consider them simply as machines which pull the carriage, and leave them wholly to the servants.

The gag bearing-rein furnishes another example of the fact that interference with one part of an animal is injurious to the other parts. When the horse is standing with its legs outstretched and its spine contracted, the weight of the body rests on the heels of the front hoofs, and not on the flat surface.

Not only that, but the horse, in addition to its weight, is exercising a muscular pressure on the heel. Thus, as was pointed out by Mr. Darby, the heel becomes inflamed and the terrible navicular disease sets in. Besides, the abnormal strain upon the tendons is so great that they give way, and the horse cannot use its limbs in the natural manner.

So we find that the gag-rein, which interferes with the neck and spine, also injures the feet. Now we will see how it injures some of the internal organs of the animal, such as the brain, heart, and lungs, and will take the authority of Mr. S. Gill, V.S., as quoted by Mr. Flower. 'Members of the veterinary profession are by no means ignorant of the various diseases produced by the use of the bearing-rein—roaring, apoplexy, coma, megrims, inflammation, and softening of the brain, all following the barbarous use of this rein.'

There is not one word to be said in favour of the gag bearing-rein. No one even pretends to aver that the gag-rein serves any purpose except that of show. Its only object is to hold the horse's head up higher than Nature intended, and as long as that purpose is served the driver is content.

One would have thought that such a machine of torture as the gag bearing-rein would have satisfied any one, but in America an addition was made to it in the form of the 'Burr' bit. Owing mostly to the efforts of Mr. Henry Bergh, of New York, it is now very seldom if ever seen. I looked out carefully during my six months' stay in the country, but never saw one. So I take the account from an American source, in an article entitled 'Henry Bergh and his Work,' in *Scribner's Monthly* for April 1879 :—

'The coachmen of the city, mostly without the knowledge of their employers, began using a round leather bit-guard, barbed with short spikes, so that when the reins were tightened the nails sunk into the side of the horse's head and made the animal exhibit a very fashionable degree of mettle.'

The reader will notice that the Burr was mostly employed by the coachman without the knowledge of the owner. But the owner had no excuse for such culpable ignorance. Mr. Bergh could discover it and insist upon its removal, and if he could do so, *a fortiori* could the owner.

## CHAPTER XIII.

The bearing-rein continued—The locomotive and the brake—Probation of an engine driver—The bearing-rein and the break—Leading reins converted into bearing-reins—Railway companies and the bearing-rein—Theories as to the bearing-rein—Its supposed use in preventing the horse from falling—Bearing-reins and hills—Harness in Scotland—The bearing-rein in Bristol—Mr. Cracknell's testimony—The bearing-rein and runaway horses—A grievous experience—The shoulder injured by the bearing-rein—Testimony of more than a hundred veterinary surgeons—'Roaring' caused by the bearing-rein—Mistaken zeal—Summary—'Free Lance's' check for a runaway horse.

If we can imagine the driver of a locomotive engine putting on the brakes and then turning on full steam, so as to produce plenty of puffing, and smoke, and sparks, and thinking that he was 'showing off' his engine to the best advantage, we should set him down as a maniac, or at the best an ignorant fool. Certain it is, that no engine owner would keep the man in his employ for five minutes.

Yet the coachman who uses the gag-rein is doing exactly the same thing. He is wasting the power of the horse in snorting, and foaming, and pawing, and head-tossing, while preventing him from drawing with his weight as he ought to do, and all the time is under the impression that he is 'showing off' the animal. The fact is, that just as fashionable women dress against each other, so do fashionable coachmen show off their horses against each other, and are afraid of each other's sneers.

Abandoning the gag bearing-rein as a torture-giving machine invented solely to minister to the pride of man, and without the least shadow of excuse, I shall try to show that any bearing-rein is doubly injurious —firstly to the horse, and secondly to its owner.

For the present, I shall put aside the question of humanity, and consider the horse simply as a convenient traction engine.

Every one will concede that the object of a traction engine is to draw its load, and that any piece of machinery which hinders it from drawing must be injurious to it. Moreover, every one possessed of common sense will concede that, in order to be profitable to the owner, the machine should be made to do all the work that can be got out of it, and that it should last as long as possible. Lastly, it is clear that the working cost should be reduced to the lowest limit consistent with the proper working of the machine; in other words, that there should be no waste of the fuel that produces the power, nor of the power when it is produced.

Now, when we have to deal with a machine, we take care that all these conditions shall be fulfilled. In a very interesting work called 'Engine-driving Life,' there is a most curious history of the ordeals through which a man has to pass even before he gains his certificate as a fireman or 'stoker,' 'an accomplishment that thousands have tried at and failed.' The book is of especial value as having been written by Mr. Michael Reynolds, a man who has gone through the whole of the ordeals himself.

There we learn how as a small boy he begins by being sent into the fire-box to take out, clean, and replace the bars and polish the tube-plate—this work often being done at a temperature of 250°. When he is too big to crawl into the fire-box, he is promoted to 'cleaner.' Now, cleaning an engine after a run occupies at least ten hours of steady work, and the boy has to learn not only how to clean, but when to clean each part—*i.e.* which parts must be cleaned while they are hot, and which can wait until they are cold.

Then he has to pass a time of probation in firing, beginning with shunting engines, and learning all he can, before he can obtain a certificate which enables him to work on a passenger train. So he goes on, always learning, his doings of every day being recorded, so that the record can be produced against or for him; and even after all this work, he has to pass a severe examination before he is allowed to take charge of an engine.

Then he goes through similar tests and trials as driver of goods trains before the lives of passengers are entrusted to him, sometimes being sent back as fireman, and having to go through all the trials again, until he works his way to the top of the tree, and is allowed to drive an express.

I wish that something of the kind could be done with the drivers of the locomotive engine which we call a horse. If authorised boards of examiners could be formed, and horse-owners would agree to employ no man or boy who could not produce a certificate showing that he was competent to do the work for which he engaged himself, we should find the lives of our horses nearly, if not quite, doubled in length, the amount of work trebled, and the cost halved. As the reader will see, I am only appealing to the personal interests of the owners, and am considering the horse as a mere machine, without more feelings than if it were made of steel and brass.

'Upon the railway,' says Mr. Reynolds, 'nothing should be regarded lightly,' and so it ought to be with regard to the horse.

One of the first faults of the bearing-rein, considered as a part of the machinery, is, that it prevents the horse from doing his full amount of work, and therefore is a cause of loss to the owner. The horse is made so as to draw by means of his weight, added to the propelling power of his hind legs. In order to do this, he must be able to lean forward, and if he be taking a load up hill, a free horse will fling himself so far forward, that his nose comes half-way to the ground. But this is all wrong to the professional eye, which likes to see a horse's head well up, no matter what work he may be doing. So the bearing-rein comes into operation, and up goes the horse's head. Consequently, instead of being able to fling his weight into the collar, the horse is obliged to scratch himself along by the muscles of the forelegs, digging his toes into the ground so as to secure a hold. The legs, not being intended for that kind of work, soon become strained, and so the horse is worn out before its time. The bearing-rein, in fact, is to the horse what the brake is to an engine. No one would be foolish enough to waste the power of an engine by keeping the brakes down during work time, and exactly similar is the result of working the horse while the bearing-rein is used. My house is close to the High Level Railway station at Upper Norwood, where vast quantities of coal, bricks, and other heavy goods are constantly delivered. There is scarcely a waggon in which the horses are not hampered with bearing-reins. Even if there be no separate bearing-reins, the leading reins are converted into them by being twisted round the hames of the collar.

A horse naturally bows its head and neck at each step, this action being intensified when it is drawing a weight up hill, and adding to the power of draught. Yet some of the horses are 'borne up' so tightly that their heads are nearly immovable, while the others are brought up sharply by the bit against the corners of their mouths each time that they try to add to their power by allowing the head and neck to droop as they step. It is quite pitiful to see the animals straining at their loads on the 'City of the Seven Hills,' as Upper Norwood is popularly called. Four horses will be seen with their heads dragged back, struggling and scratching to pull up the hill a load which would not be too much for three similar horses with their heads free. Reasoning and remonstrance are equally thrown away upon the drivers. They have been used to the bearing-rein all their lives, and cannot be made to believe that they may possibly have been wrong all their lives. If an ordinary carter be told that several railway companies employ some two thousand horses each in similarly heavy work, and never allow the bearing-rein to be used, not the least impression will be made on his mind; neither will he be convinced if he be shown the magnificent horses belonging to many great London firms, such, for example, as Messrs. Barclay and Perkins, all of them doing their work with free heads.

Even the owners cannot depend on the men. The strictest orders may be given, and may be obeyed as long as the men can be watched, but no sooner is the master's eye withdrawn than on go the bearing-reins again. It is easy enough to take away the bearing-reins themselves, but the horses must have reins of some sort, and these can always be converted into bearing-reins, as has already been mentioned.

I fear that the only cure will be that which Messrs. Barclay and Perkins have employed so successfully—*i.e.* instant dismissal on disobedience to orders.

Sometimes the personal argument will be effectual, at all events for a time. Give the man a light cart, put him between the shafts, and set him to pull the cart up a hill. He will instinctively lean forward in proportion to the heaviness of the task. Then set him to perform the same work, but do not let him lean forward.

If this little experiment be conducted with good humour and tact, the man will mostly be convinced.

But he will probably be so ridiculed by his comrades that in self-defence he will revert to the bearing-rein.

. Now we come to the theory that the bearing-rein prevents the horse from falling.

A more ludicrously absurd idea never entered into a human head, and it is wonderful that a child of six years old should believe that tying a horse's head to its tail or its collar can prevent it from falling if it stumbles.

Let anyone who believes this to be the fact tie his head back to his waist by a similar strap, and then run over rough ground. He will find himself even more likely to fall than when his head was free.

If the bearing-rein were fastened, not to the horse, but to the top of the carriage, there might be some show of reason in the idea. Even in that case the horse is so heavy that no rein could hold it up, but must snap under the strain. Supposing that it could sustain the horse, matters would be still worse for the animal, which would be left with its whole weight hanging on its mouth. What mostly happens when a tightly ' borne-up ' horse stumbles, is that he falls forward, and the rein is either snapped, or the hook is pulled out of the saddle.

Should the horse fall, it would be almost impossible for the horse to recover its feet unless the bearing-rein be removed. In fact, as Mr. S. Gill very rightly says, 'To imagine the animal is prevented from falling by this reining up of the head is an error. The pressure on the veins and arteries impeding the flow of blood, it is impossible for the animal stumbling to recover himself.'

It is a remarkable fact in the history of the bearing-rein that in England hilly ground is always made an excuse for the employment of this rein. Here, close to the Crystal Palace, the absence of level ground is always brought forward when anyone tries to persuade a driver to remove the bearing-rein.

It is not so in Scotland, and yet Scotland is far more hilly than England. Any of my readers who know Edinburgh, for example, and have been obliged to walk from the New Town to the Castle against time, must be tolerable judges of a hill. Yet during the whole time that I was in Edinburgh I saw scarcely a bearing-rein in use. The very few that I did see belonged to gentlemen's horses, and had been imported from England.

The Scotch are far too canny to act so foolishly. They know the value of a horse's work, and do not choose to hamper the animal with a contrivance which prevents it from putting forth its full strength.

They also know that, according to Professor Fleming, the bearing-rein 'does not prevent stumbling, but, on the contrary, predisposes the horse to fall, and with much more severity than if it were not used.' Lastly, putting the welfare of the horse out of the question, the bearing-rein adds to the expense and weight of harness, and gives additional trouble in keeping it clean. There are some firms which do not include the bearing-rein in their estimates of harness, and, if the buyer insists upon having it, a separate charge is made for it.

A few years ago I had the pleasure of giving a lecture on the subject at Bristol. I think that nearly, if not quite, four hundred carters and other drivers were present. Bristol being on very hilly ground, the bearing-rein was almost universally employed. However, after much discussion, many of the drivers did see the mistaken ideas under which they had been labouring, and in order to keep each other in countenance, were formed into a society, the members of which wore a little rosette, by which they could be distinguished.

I was rather amused with one of the men. After having abandoned the bearing-rein for a few days, he was so delighted with the result, that he procured two additional rosettes, and fastened one on each side of the horse's head.

A most valuable contribution to the literature of the bearing-rein is afforded by Mr. E. Cracknell, the well-known coachman on the London and Birmingham road, in a letter to Mr. Flower:—

'I drove the "Tantivy" coach nearly twenty years without a bearing-rein, and seldom had a horse fall, although they went at a great pace, and I frequently drove as many as seventy-two per day. The class of horses I had to drive were difficult, many of them being old steeple-chasers, hunters, Newmarket weeds, &c.

'My first experience in dispensing with the bearing-rein was between Henley-on-Thames and Henley; it was the practice to walk the horses, the greater part of the hill being very steep. One day I left off the bearing-reins, and was astonished at the result; the horses never attempted to slacken their pace, but trotted the whole distance up the hill. From that time I dispensed with the bearing-reins entirely.

'There was a strong prejudice at first with my colleagues against it, but eventually they adopted my plan. I had most troublesome, dissipated horses to manage, but *with a light hand and their heads at liberty* they generally became tractable.'

The reader will observe that Mr. Cracknell does not state that horses never fall when the bearing-rein is not used. When seventy horses are driven daily at a fast pace, it is not reasonable to expect that there should be no falls. He only states that the falls were few, and in another part of the letter he refers to the acknowledged fact that if a horse falls when he is wearing the bearing-rein, he cannot rise until it is removed.

Putting aside the cab-horses, which, fortunately for them, are not hampered with these appliances, every other horse at least that falls in London is wearing the bearing-rein or its equivalent. Now I do not say that these horses fell because they wore the bearing-rein, and it would be equally foolish to say that the others fell because they did not wear it.

Supposing that we grant—which I do not—that the chances of falling are equal, at all events, the horse which does not wear the bearing-rein has the better chance of getting up again.

The last argument which the advocates of the bearing-rein offer is, that it is a safeguard against horses running away—*i.e.* that a horse cannot run away while it wears a tight bearing-rein. I never could see how this effect could be produced, although it is assumed as an axiom by the defenders of this rein. But it is similarly assumed that the bearing-rein protects the horse against stumbling or falling.

In fact, the pain which is produced by it will

make a horse unmanageable and likely to run away. Here is a case in point.

The late Mr. B. Shaw, M.P., possessed a horse which was so violent that the coachman ordered a very severe bit to be made expressly for it. Before the bit was sent from the maker, a groom was exercising the horse, and, in order to prevent it from running away and to keep it under control, the bearing-rein was employed as usual, and drawn back up to the last hole.

The horse was, as the groom said, exceedingly 'fresh' and restive, and struggled so violently that the rein broke. 'I thought it was all up with me,' said the groom, 'but to my surprise the horse became at once manageable, and went beautifully, and the severe bit was never used.'

Lord Palmerston used to say that 'a runaway horse is best kept in by a light hand and an easy snaffle,' and his opinion is corroborated by that of Mr. Cracknell, as quoted above.

The best authorities are unanimous in their condemnation of the bearing-rein. Sir Francis Head, who did as much miscellaneous horsemanship as man can well do, always employed the strongest language against it.

The authors of standard works on the subject equally condemn it. Mayhew never loses an opportunity of denouncing it. In classifying horses which are crippled by injudicious treatment, he proceeds as follows: 'A fourth set are rendered cripples by the unfeeling use of the bearing-rein, which disables the organs of respiration, and renders the lightest draught a terrible burden, by throwing the work on the muscles of the limbs, while it compels these agents to contract at a terrible disadvantage.

'Those who delight in a lofty crest may accomplish more by attention to the health and diet than by the absence of humanity. The strongest bearing-rein and the sharpest bit cannot exalt the head of a spiritless horse.'

In another part of the work he shows by a series of diagrams how the best shoulder is ruined by the bearing-rein, and how the horse, from having its head raised abnormally and prevented from seeing the ground, by degrees acquires a step which is almost identical with that of a blind animal.

Youatt, in his well-known work, writes equally strongly on the subject, and so does Fleming.

If any other argument could be required, it may be found in an important document which was issued a few years ago:—

'We, the undersigned, are of opinion that the use of the bearing-rein, when tightly applied, is painful and irritating to horses, is directly or indirectly productive of disease when regularly worn, and by its mechanical action greatly hinders horses from employing their full strength. For the above reasons—on the plea of utility as well as humanity—its use should be discontinued.'

This document is signed, not by 'theorists' or 'humanitarians,' nor by ignorant and impulsive women, but by upwards of a hundred well-known veterinary surgeons, six of whom are professors in the veterinary colleges of London, Edinburgh, Glasgow, &c. Twenty-four of them are Fellows of the Royal College of Veterinary Surgeons, and the remainder are Members of the College.

Several accompanied their signatures with additional remarks.

Professor Axe, of the Royal Veterinary College, London, makes the following statement: 'Eleven years' experience in the post-mortem house and the dissecting-room of our college has made me acquainted with a variety of structural alterations and deformities arising from this cause, and which must have rendered life a burden *and shortened its span.* . . . If the public could see and understand the effects of its insidious work on the respiratory and other organs, I do not think that its use would be long continued by them.'

Mr. W. G. Taylor, of Nottingham, in allusion to the adjective 'tight' as applied to these reins, adds that *he disapproves of the bearing-rein in any form.*

Mr. J. V. Blake, of Ryde, states that many horses have come under his notice with their tongues partially severed by its use.

Mr. W. R. Marriott, of Colgrave, Nottingham, cites it as one of the causes of 'poll-evil'—*i.e.* abscess on the head; and Mr. J. Freeman, of Hull, F.R.C.V.S., writes as follows : ' I have often been asked for something to rub the glands of the throat with when a tight rein has been used In one case of injury by the bearing-rein, I begged the driver to leave it with me for three weeks. At the end of that time he called, saying that his horse had got quite well without medicine.

That the bearing-rein causes ' roaring ' and other diseases of the respiratory system we have already seen. The cause is simple enough. The windpipe is forced into an unnatural curve by the bearing-rein, and the supply of air is therefore checked in its passage to and from the lungs.

As to the limitation which is apparently implied by the words 'tight bearing-rein,' it is really no limitation. A bearing-rein which does not ' bear ' on the horse—*i.e.* is not tight—is practically no bearing-rein at all, and only a piece of supplementary harness hung at the horse's head by way of ornament.

Absurd as it may seem, I have personally known more than one case where the owner compromised with his coachman, and permitted him to retain the bearing-rein provided that it were too slack to press against the mouth even if the horse fell. The man was afraid to meet the professional eye of his fellows, and was not proof against the sneers to which he knew he would be subjected if his horses wore no bearing-reins.

One of the most important points in the veterinary surgeon's overwhelming condemnation of the bearing-rein may not have struck the reader. It is the strong sense of honour shown by the writers.

Every reform has some adherents, whose zeal outruns their discretion, and who do more harm to their cause than if they had been active opponents. For example, soon after the Battle of the Bearing-reins began, some ladies made an outcry against the neighbouring carters, accusing them of employing the cruel bearing-reins, &c. The men, however, proved that they had no bearing-reins, but only the leading reins, and that their accusers did not know the difference. Consequently, the enemy was routed.

I very much fear, however, that although the ladies might not have been able to distinguish a bearing-rein from a leading-rein, the men may after all have outwitted them, the leading-rein being immediately convertible into a true bearing-rein by a twist round the hames.

Other 'impetuosities,' to borrow Charles Reade's term, made an onslaught on the veterinary surgeons as a body, accusing them of encouraging, or at least conniving at the use of the bearing-rein, because the horses were injured by it, and so grist was brought to their mill. I use, as far as I can remember, the identical terms.

Now, in this indictment there was just that scintillula of truth which makes such an accusation doubly exasperating. It is perfectly true that a large portion of the ailments of horses is caused by the bearing-rein, and that if it were to be wholly abolished, the veterinary surgeons would have fewer patients. But A has no right to assume that because B would gain by perpetrating a dishonourable action, B would do it. Still less has A a right to assume that B actually does perpetrate that action, and still less to accuse him of it publicly.

The above manifesto of the veterinary surgeons is a complete answer to the charge, and shows that, although by advising the use of the bearing-rein they would sensibly increase their gains, they are far too honourable to do so at the expense of the horse.

Now suppose that we sum up the arguments in favour of the bearing-rein and against it.

On the one side, the bearing-rein is said to impart an appearance of mettle to the horse, and to make it look imposing. It is said also to save the horse from stumbling or falling, and to prevent it from running away. If these statements could be proved, no owner of a horse would be justified in refusing to confer such a benefit upon his horse. But how does the case really stand?

In the first place, the appearance of the horse is *not* improved, except to the professional eye, which, not many years ago, demanded that the ears should be cropped and the tail docked and 'nicked' before that eye's requirements were satisfied. To the eye of the naturalist or artist the artificially distorted attitude of the animal becomes a deformity. 'Imagine,' writes a correspondent of the 'Animal World,' 'a Landseer being asked to paint one of these deformed fore-quarters and straddling bodies while suffering from the gag-rein!'

Secondly, we have seen that if the bearing-rein has any influence upon the tendency of the horse to fall, it rather increases than decreases that tendency; while the assertion that it holds up a horse from falling, by tying his head to his back, is so glaringly absurd as to require no refutation.

SUMMARY. 241

Lastly, it has been shown that bearing-reins and severe bits cannot prevent a horse from running away, nor stop him when he has started. On the contrary, it has been shown that they are much more likely to incite him to run away.

Now let us see what the opponents of the bearing-rein have to say on their side of the question. In the first place they have proved that the bearing-rein causes torture to the horse in proportion to the tightness with which it is drawn. They have proved beyond the power of contradiction that it causes various diseases and ailments, roaring, poll-evil, apoplexy, spavin, navicular disease, swelling of the glands of the neck, and distortion of the shoulder. That it deprives the horse of its powers of traction in proportion to the tightness of the rein no one has attempted to deny. That it 'renders the life of the horse a burden and shortens its span,' has been shown by Professor Axe. The injury which it does to the sensitive mouth of the horse is too evident to need proof.

Added to this, we have the unanimous condemnation of it by more than a hundred of our best veterinary surgeons.

Even granting that it did give the horse a more mettlesome aspect, that it did keep the animal from falling, and that it did prevent him from running away, we should have no excuse for attaching to the animal an appliance which keeps him in constant pain while he is at work, which inflicts upon him a variety of painful ailments, and which helps to wear him out in a life of suffering before he has passed the half of his natural existence.

PARENTHETICALLY, with regard to the runaway horse, Mr. G. Ransom ('Free Lance') has invented a contrivance which, without inflicting the least pain, will stop the most infuriated horse within a few yards. He was good enough to present me with his original model. It is made of brass, and somewhat resembles a lady's hinged bracelet not quite three quarters of an inch in width.

When fitted, it passes over the nose and is not in the least conspicuous. If the horse attempts to run away, the driver or rider has only to pull a strap connected with the apparatus, which has the effect of closing the nostrils. Now the horse, when its nostrils are closed, is quite helpless, not being able to breathe, and so is brought to a standstill.

The North American Indian women have a similar mode of teaching their babies not to cry. The first time that a baby raises its voice, the mother closes its mouth and compresses its nostrils, and as the process is repeated whenever the child cries, it very soon learns to be silent. In consequence of this treatment, such a sound as the crying of a baby is never heard in a native encampment, even though there may be plenty of babies in it, lying on the ground, or hung up to branches to be out of the way.

Perhaps Mr. Ransom, who knows the Indians well, took his idea from this custom.

## CHAPTER XIV.

The BLINKER and its supposed uses—Cropping of ears—Alleged necessity for cropping—Contradictory temperament of the horse—Courage and timidity—Inquisitiveness—Rarey's principle—The kettle-drummer's horse—Mr. C. H. Tamplin's experience—Obstinacy of a coachman—Value of the 'master's eye'—Waterton at Walton Hall—Letter from 'C. F. W.' to the *Field* newspaper—'Jockeying' adviser—Modified blinkers—Eye of the horse—Cruel superstitions—The third eyelid, or 'nictitating membrane' and its use—The groom's rashness and its effects.

WE have not quite finished with the horse's head.

Attached to the head harness of horses used for driving are almost invariably a pair of leather flaps called blinkers, which cover the eyes, and prevent the animal from seeing objects behind or on either side of him.

For riding purposes, blinkers are never used, or at all events so seldom, that if a ridden horse were to be seen with them, considerable astonishment would be excited, and in all probability the horse and its rider would have to endure no small amount of derision. In America these appliances are called blinders.

Their object is to prevent the horse from seeing objects which will frighten it. Why a horse should be frightened when it is driven, and not frightened when it is ridden, is rather difficult of comprehension. If we accept the assumption that a horse must be frightened at objects which it can see, we ought to be consistent and assume that it must be frightened at sounds which it can hear. And, to be consistent, we ought to stop the horse's ears as well as blind its eyes. Indeed, seeing what pranks are played with the horse, I very much wonder that ear-stoppers have not come into fashion long ago. Perhaps they might have done so, if they could have been adorned with a Greek or Latin title, such as Otoclids or Auriclauders, made ornamental, and, like blinkers, could bear the crest of the owner. There are many men still living who can remember when a horse was considered quite unfit to be looked at unless his ears were cropped close to his head, just as was the case only a few years ago with many breeds of dogs.

At the present day we should say that the whole beauty of the head was destroyed by the loss of the mobile ears, which indicate the emotions which pass through the animal's mind, and that the horse was hopelessly disfigured. We might also say that to crop the horse's ears was indirectly dangerous to man, because the mode of carrying the ears is one of the tests whereby to judge a horse's temper.

All this is perfectly true. But it was equally true in the days when cropping was in fashion, and yet its truth had not the least effect on the advocates of the custom.

This figure of the Cropped Horse is taken from

A 'CROPPED' HORSE.

a hunting print of the period in which cropping was in fashion.

At the present day the ear is, happily for the horse, allowed to retain its full dimensions, so that the animal can direct it as he chooses, and be sensible, as he ought to be, to the slightest sound. But, though the groom may not cut the ear off, he cannot let it alone. The inside of the ear is furnished with a supply of hairs so arranged that they exclude dust, flies, and other extraneous objects, while they do not interfere with the passage of sound.

Therefore, the presence of these hairs is an abomination to the professional eye, and the groom, if left to himself, will remove the hairs as far as possible, sometimes cutting them off, and sometimes, when he wants to be very professional, singeing them down with a candle flame. Deafness is often caused by this atrocious practice; but that is nothing to the groom as long as the horse is got up according to the prevailing fashion.

If the reader will again turn to the horse 'Fashion,' on p. 200, he will see that I have ventured to add a pair of ear-stoppers to the blinkers, the gag bearing-rein, the hogged mane, the cropped ears, and the docked and nicked tail, all of which are, or have been, considered necessary to the gratification of the professional eye.

The real fact is, that blinkers are rather the causes of fright than the preventives, and for this reason. The temperament of the horse is most peculiar, and especially as regards courage. To judge him from one point of view, he seems an arrant coward. To judge him from another, he is one of the most courageous animals on the face of the earth. You may see him scared by a feather blown against his nose, or by the sound of a boy's popgun, and would naturally consider him to be a coward.

But the very animal that was frightened at a feather, when properly taught will walk about unconcernedly among locomotive engines, caring nought for their puffings and snortings, and whistlings, and 'blowing-off' of steam.

The very animal which ran away at the sound of the popgun will, when properly trained, face a square of infantry despite the glitter of the bayonets and the flash and rattle of the musketry. He will undauntedly charge a battery, and so from this point of view he must be reckoned as exceptionally courageous.

'He mocketh at fear, and is not affrighted; neither turneth he back from the sword.

'The quiver rattleth against him, the glittering spear and the shield.

'He swalloweth the ground with fierceness and rage; neither believeth he that it is the sound of the trumpet.

'He saith among the trumpets, Ha, Ha; and he smelleth the battle afar off, the thunder of the captains and the shouting.'

This magnificent description is as true an eulogium of the horse's courage as it was when penned by the unknown writer of the Book of Job, at least three thousand years ago.

Like ourselves, the horse fears the unknown, or rather the imperfectly known, and as soon as he becomes familiar with the dreaded object, his fears vanish. It was solely by acting on this temperament that Rarey was so successful in the management of troublesome horses. He went on the principle that the 'vices' of the horse are almost wholly due to his fears, which have not been understood by man.

Nothing could be simpler than Rarey's system, of which I have often been an eye-witness. He would take any object of terror, say a revolver, or a drum, or an umbrella, and manage by degrees to make the horse understand that it would not hurt him.

Then he acted on another characteristic of the horse.

The animal is as inquisitive as a cat, so inquisitive indeed, that no matter how much an unknown object frightens him, he cannot keep himself away from it. He will stand at a distance gazing at it with outstretched neck, and sniffing as if the sense of smell could tell him something about it. Perhaps a new access of fright will take place and he will gallop off, but in a short time is sure to come back again, drawn as if by some magnetic attraction.

Thus he will go on, always drawing nearer and

nearer until he is close to the dreaded object. Then he will smell it, test it by rubbing it with his nose, which is the tactile organ of the horse as the hand is with us, and then, having satisfied his curiosity, will trouble himself no more about the former object of his fears.

The following short account of Rarey's dealings with a horse which was too much even for the Life Guards, is taken from Mr. S. Sidney's 'Book of the Horse:'—' The commanding officer of one of the regiments of Household Cavalry placed in Rarey's hands a cream-coloured stallion from Her Majesty's stud that had resisted every effort of roughriders of the regiment to make it carry the kettle-drums.

'The horse was in the riding school. Rarey took one of the drums, placed it before the horse's nose, and by degrees got him to smell it. Then he gave it a slight tap with one of his fingers, on which the horse started, but smelt it again. Repeating this operation again and again louder, and each time with a drumstick, after a series of starts and smells the horse began to find out that the drum did him no harm.

'The drum was then placed against his side and the tapping process repeated. Finally, certainly within less than an hour, Rarey mounted his pupil and marched him round the school, beating the drum loudly. From that time forward the cream stallion bore the gorgeously attired drummer, beating the silver kettle-drums, and pacing proudly at the head of the regiment.'

I have several times seen him perform not only a similar feat, which is a comparatively easy one, but in less than half-an-hour teach a shying horse to allow all the chambers of a revolver to be fired round his head, a flag to be waved violently before his eyes, and an umbrella to be opened suddenly in his face.

I fully believe that if the owner of a horse were to begin driving it without blinkers, it would never require them. But of course I admit that if a horse which had always been driven in blinkers were to be suddenly deprived of them, he might probably take fright at the unaccustomed range of vision. This feeling, however, would soon wear off, and then the horse would not only have wider opportunities of finding out the meaning of objects which would have frightened him if half seen, but would look all the better for not having his beautiful eyes concealed behind the leathern flaps.

Still it would be quite possible to accustom the horse to do without them. At first, they could be spread out, so as to allow the animal a wider range of vision. Then they could be gradually lessened in size, and lastly removed altogether.

People often say to me, 'We do not deny that horses might do well enough without blinkers as long as they keep to country roads, but if they are driven in London, or any other populous city where there is much traffic, they would be frightened at the other vehicles, and especially at the whips of all the drivers.' I suppose that Londoners think just the same of the country, and that the horses would be afraid of the trees and the wide expanse of country.

If people who owned horses blindfolded them entirely before driving them, there might be some sense in this idea; but as even the tightest blinkered horse can see in front, and can therefore see all the whips which are likely to hit him, even this superficial argument fails. Of course, when a country horse is just brought to London, it is nervous and alarmed at the unaccustomed sights and sounds, and so is a London horse when taken into the country, whether it be blinkered or not. The blinker has nothing to do with the question.

In a letter addressed to the 'Animal World,' as long ago as 1873, Mr. C. H. Tamplin, a London surgeon, narrated his experiences with the 'blinkers:'

'Two years ago the idea occurred to me to try, and if possible find out, the reputed use of blinkers. Our coachman was ordered to come round without them. He did so (bearing-reins we had renounced as instruments of torture long before), and the horses, which had always been used to blinkers before, went along beautifully, and evidently enjoying much more comfort. In the afternoon we had another pair harnessed in the same way, and the result was precisely similar.

'To replace a horse which had died, another was bought. He also behaved extremely well without blinkers or bearing-reins. Two others in the country, one a mare which used to shy and jib inveterately, and the other, almost thoroughbred, went as quietly as possible without blinkers, those appendages having been removed from their harness.

'Much stranger still, a young Irish horse, five years old, almost thoroughbred, had never been driven in double harness before, and had never been taken into any town in his life until he came to London, where he was ridden for two weeks that he might get used to the traffic.

'At the end of that time our coachman, *without directions*, put him into double harness with blinkers. He went very well until a large dog ran across the road before him, when he shied and nearly jumped over the pole. He would probably have taken to kicking had not some one rushed up to his head *and pulled the blinkers aside that he might see the dog*, when he became absolutely quiet.

'The next day our coachman removed the blinkers, and from that time he has behaved admirably amidst the busy traffic of the streets, without evincing a tendency to shy.

'The inference I draw from this and other cases is, that blinkers are totally unnecessary, and do more harm than good. In passing other vehicles at crossings and when turning, instead of going blindly on at the risk of running against whatever may be near them, horses without blinkers know what is wanted, and so become much more easy to drive.'

The reader will probably have gathered from this narrative the necessity for the master's eye as the only check upon the perverse obstinacy of servants, who will always revert to their accustomed ways whenever they can find an opportunity. It cannot be too often repeated that argument with a servant is useless, and should never be used.

'You may prove to your coachman and grooms that scooping out the interior of the hoof is destruction to the horse; that the bearing-rein causes horses to fall, and that the blinkers induce them to shy. Your reasoning will " go in at one ear and out of the other," without producing the least impression on the man's mind; and, unless he knows that he is closely watched, he will carry off the horse to the farrier's at the first opportunity and have the frog and bars cut off and the heel opened. He will replace the blinkers and fasten the horse's head back with a bearing-rein, thinking all the while that his master is a fool, and that his own way is the only one for a horse.'

The late Charles Waterton afforded a valuable example of the value of the master's eye.

When he came into possession of Walton Hall, he determined to give Nature a fair trial against long-established customs. A keen preserver of game, he would not allow his keepers to interfere with the hawks, rooks, crows, magpies, starlings, stoats, weasels, and other 'vermin.' He had to run counter to the prejudices of every man on his estate, not to mention all his neighbours, and soon found out that he must never argue, but give his orders and see that they were obeyed. Every one in his employ knew that every part of his work would come under 'the Squire's' eye in the course of twenty-four hours, but when the master would make his appearance no one knew, except that it was sure to be when he was least expected.

One of his rules was, that after the day's work the horses should not be tied up, but allowed to walk about at liberty in the stable-yard. One night he suddenly asked me if I would go into the grounds with him. So he lighted a lantern, and we went over a large portion of the premises, finishing with the stables.

In one of them a horse was tied up. Waterton cut the halter into little bits, laid them on the floor, and said nothing. About five next morning the bailiff passed a very unpleasant quarter of an hour. 'Please zur, I thowt,' began the bailiff. 'You *thowt*! You thowt! Nobody thinks here but me,' returned Waterton, and with a stern warning sent him out of the room, not letting him say another word.

The following letter, addressed to the 'Field' newspaper of June 20, 1877, put the bearing-rein and blinker question in a very amusing but forcible summary.

'Sir,—I think there is a great deal of nonsense written about this bearing-rein business.

'The facts of the case seem to be as follows:

'Horses are naturally very stupid animals; they can neither carry their heads nor their tails in a proper manner, neither can they make a proper use of their eyes, therefore it is necessary that they should be taught the proper use of them.

'Now a horse naturally wishes to see the ground in order that he may not tread on anything likely to make him tumble down. But we make our roads so smooth that there is no necessity for his seeing them, so we very properly tie his head up to make him look at the sky, which gives him a very elegant appearance, especially when he gets used to it and takes naturally to what is called " star-gazing."

'But even then, he might look to the right or left, and so see something or other which it is far better for him not to see, therefore we half blind him. This is a great point, for with his head tied up, and his blinkers on, he cannot possibly see anything but the sky, where there is nothing to see, and consequently nothing to frighten him.

'Sometimes the brute is badly made, being small in the chest, in which case the saddle, or whatever it may be termed to which this bearing-rein is fastened, is pulled forward by it, in which case it becomes necessary to have a crupper to hold the saddle back, and thus in a manner you tie his head and tail together, so that he cannot put down his head without pulling up his tail.

'I think the above letter gives ample reason for the use of the bearing-rein and blinkers.

'I am so convinced that this is the right and proper way to educate a horse, that I mean, when I can afford it, to have a horse to ride; and as I suspect I shall not be a very accomplished horseman, I mean to have a hook at the front of the saddle on which to put a good tight bearing-rein—this is to keep him

S

from running away with me. I shall also put on blinkers, so that he can see neither to the right nor left, and think that under these circumstances I shall be tolerably safe.—C. F. W.'

There is only one argument which is in the slightest degree in favour of the blinker. Among horses and among mankind there are sure to be some who have the strongest disinclination to do any work that they can get any one else to do for them; and with them, as with us, this inclination becomes stronger as they grow older and more artful. So, when a pair are driven, one being a willing horse and the other a slug, the latter will shirk his share of the work, and leave it to his companion.

I saw just such a pair while I was in Boston, Mass. They belonged to one of my friends, and as we were driving, I asked him why he put blinkers on his horses, and why those blinkers were quite unlike the usual form. He said that he was as much opposed to the blinker as myself, and would not use it until compelled by the laziness and cunning of one of the horses.

The animal could pull quite as well as his companion, but he was a slug and a swindler. He always contrived to keep his eye on his master, and as long as he saw that he was watched, he did his fair share of the work; but, if the driver's attention were diverted in any way, the horse dropped back from the collar, but used all the action of a pulling horse, so as to deceive the driver. Consequently, his comrade—a very willing horse—had his work nearly doubled.

As soon as the impostor perceived signs of the driver's renewed attention, he went forward into the collar, and really pulled instead of making believe. He did it so cleverly that several times, before the trick was found out, the whip came on the wrong horse. In order to counteract this tendency, his owner had a special set of blinkers made. They are scarcely half the size of the ordinary blinker, and they stand out nearly at right angles from the head, so that the horse can see everything in front or at the sides, but not behind.

This simple arrangement completely checkmated the horse. He tried his old ruse once or twice, but always found himself reminded by the whip that he must play no more tricks. For the sake of symmetry, both horses wore similar blinkers, and when I saw them, it was impossible to tell which had been the delinquent.

Under no circumstances, however, ought horses to wear the large blinkers which are occasionally seen.

Sometimes the front edges of these blinkers are

drawn so closely together that there is only a narrow strip of vision in front, and none on either side. These close blinkers have two bad effects on the eye. In the first place, they heat the eye-ball by preventing the free access of air to it; and in the next, by forcing the animal to keep its eyes always directed forwards, they fatigue and strain the delicate muscles by which the eye is moved.

In this place it is very tempting to describe the structure of the horse's eye, but such a description would be outside the province of this book, which only deals with those structures of the animal which ought to be understood by those who have the charge of it.

Considering the manner in which the horse is treated by man, I almost wonder that fashion did not recommend the entire instead of the partial blinding of the horse. Had fashion turned in that direction, no question of humanity would have stood in its way. Those who crop the ears of horses or dogs, or who cut off their tails at the demand of fashion, would have no scruple in going a step farther, and putting out the horse's eyes. The reader may think that I am going too far in making such a statement, but I am simply stating the truth. All who know anything of 'fancy' are familiar with the fact that it is the custom to put out the eyes of certain birds, because it is thought that their attention, being shut off from distraction by external objects, would be concentrated in their song. Some birds escape total blindness by being 'scaled.' A red-hot knitting needle is held so close to the eye, without actually touching it, that a white 'scale' is formed on the cornea, and prevents light from having access to the eye until the scale is thrown off, and the cornea restored by the reparative operations of Nature. Many, however, if not most, are totally blinded, the red-hot needle being pushed into the pupil of the eye. This is no modern practice, as is shown by one of the photos in Hogarth's 'Progress of Cruelty.'

The custom, which still lingers, of slitting the tongue of a starling or magpie, in order to enable it to speak, is scarcely less cruel. Oddly enough, its gradual extinction has been owing, not to any increase of humanity on the part of man, but to the improvement of our coinage.

To slit the tongue with steel was thought to be useless, and the only instrument which possessed the requisite virtue was a 'silver sixpence.' Some of my readers may be old enough to recollect the sixpences which were current in my childhood—mere irregular discs of metal, with scarcely a vestige of obverse and reverse, and worn at the edges until they were as sharp as knife-blades. The doubly thick 'milled' edge of the modern coinage, however, has had the curious effect of saving from a painful operation thousands of starlings, jackdaws, and magpies.

When I was at school at Ashbourne, in Derbyshire, I remember one old sixpence, which belonged to a man of sporting propensities, and which was kept sharp expressly for the purpose of cutting birds' tongues.

Now, I maintain that the men who will burn out the eyes of a bird to make it sing, or slit its tongue to make it talk, or cut off the ears or tail of a dog or a horse at the demand of fashion, would not hesitate to go one step farther, and put out the eyes of the horse if the prevailing fashion required that it should be blinded.

Fortunately, fashion has not as yet gone so far, and the groom may not destroy the eyes of the horse. But, in order to be in the fashion, he does go as far as he can towards injuring the eye.

In the first place, the eyelashes, which, of course, are intended for a protection for the eye of the horse as for that of man, offer a temptation which the fashionable groom cannot resist, and he cuts them off. The eyelashes will, however, grow again, and the harm is but temporary. Far different, however, is it with another structure connected with the eye.

In common with many mammals, with birds, frogs, and several other vertebrates, the horse is furnished with a sort of third eyelid, or 'nictitating membrane,' as it is scientifically termed. In the horse and dog it goes by the popular name of the 'haw.' Without closing the external eyelid, the animal can draw the nictitating membrane over the eye, so as to sweep off any dust or extraneous substance that might injure the eye. This movement may be seen to perfection in the owl tribe.

It sometimes happens that this membrane becomes inflamed, mostly, I believe, from the ammoniacal vapour of ill-ventilated stables, and then it is apt to protrude from the corner of the eye. The groom, who naturally dislikes the red, unsightly projection, never thinks of treating it as a case of subduable inflammation, but cuts it off. From that time, the membrane can never do its work rightly, and the consequence too often is a partial or entire blinding of the eye.

## CHAPTER XV.

The mane and the practice of 'hogging' —The tail and its office—What man does to the tail—Docking—A puzzled J.P.—The professional eye again—Docking and lock-jaw—Nicking—An unexpected ally—'Conducive to human safety'—The tail and the crupper—Winter and summer coats of the horse—Clipping and singeing—American horses in winter—Fashion with man and horse—The groom's real reason for clipping—Mayhew and Lupton's opinions of clipping—The 'moulting' of birds.

HAVING now seen some of the effects of the professional eye upon the head of the horse, we will pass to its neck and spine.

The chief glory of the horse's neck is the Mane, which 'clothes his neck with thunder.' It is balanced by the full and flowing tail which adorns the last vertebræ of the spine, both of these ornaments being distinguishing marks between the horse and the ass. It could not be expected that the professional eye could miss such splendid opportunities as are presented by the mane and tail, on both of which man has laid his sacrilegious hands.

The Creator has made the mane full and flowing to match the tail. Man, therefore, cuts it away until only some two inches are left standing perpendicularly, so as to make the neck of the horse look as much as possible like that of the ass.

This custom of cutting the mane, or 'hogging' it, as this particular form of mutilation is called, was, as far as I can discover, introduced in the early part of the century, when a sham classical mania reigned paramount in the fashionable world. Sporting men

HORSE WITH 'HOGGED' MANE.

were called 'Corinthians,' horse races were dignified by the name of 'Olympian Games,' and so forth. Therefore, the horses had to look classical in order to preserve consistency, and their manes were cut so as to make them resemble the horses of antiquity.

Lately, I regret to say that the mutilation of the mane has been revived, chiefly, I believe, by polo players.

Except that 'hogging' the mane is a shameful disfigurement of the horse, it does no harm. But when man deals with the tail, a very different verdict must be given. The tail, as the Creator made it, is shown in the figure called 'Nature,' on p. 200. One of its uses is obviously to act as a whisk, and drive away the flies which irritate the sensitive skin of the animal, especially in hot weather; and that it is an ornamental appendage, without which the form of the horse would be incomplete, is evident to any but the professional eye.

TAIL DOCKED AND NICKED.

Some years ago, while engaged in making out the heads of this work, I gathered together some notes under the title of 'Exploded Fashions.' Those which refer to the tail are as follows: 'Bang-tail, cock-tail, dock-tail, nick-tail.' With the greatest regret I have seen these exploded fashions creeping slowly but surely back again. The tail, which some years ago was allowed to preserve its natural and beautiful form, has been gradually shortened.

The result comes from the rivalry of grooms and coachmen. While driving, especially in places of fashionable resort, they naturally inspect the horses belonging to the carriages which they meet. Mr. Brown's coachman sees that Lord Kennaquhair's horses have had their tails cut short, so nothing will serve him but to cut his own horses' tails still shorter.

If this rivalry in tail-cutting extended only to the hair it would do no very great harm. It would spoil the appearance of the horse, but would not inflict pain or affect health, and the hair, if allowed to grow without hindrance, would, in process of time, resume its natural appearance.

But this tail-cutting mania goes farther, and not only the hair, but the tail on which it grew is cut off, several of the last vertebræ being amputated. The pain caused by this operation is very great, but, as we shall see, the torture is not all. How painful the operation is, is shown in the following letter:—

'IS DOCKING HORSES' TAILS CRUEL IN A LEGAL SENSE?

'*To the Editor of the "Daily News."*'

' SIR,—I am a country magistrate, who is anxious to obtain an opinion founded on the judgment of my brother magistrates throughout the country generally on the subject of this letter. My personal dislike of anything should not make me give a decision which could not be maintained if appealed against.

'I am fifty-seven years old, and from boyhood in the hunting-field, on the race-course, and in a cavalry regiment, I have had horses as my friends and servants, and I have no hesitation in deciding the practice of docking to be unnecessary and cruel. Unfortunately it is a common custom, and to be consistent, thousands should be summoned, and not, as now, only one occasionally. I think it is shown to be unnecessary for safety in driving, because when a few years ago it was not the fashion to dock horses' tails, carriage accidents were not more frequent than now.

'That it is cruel I have no doubt. Providence gave the horse a tail for good reasons, as any one who has watched a colt at grass can see. The absence of it when flies are troublesome causes absolute misery. It will also be remarked that those parts over which the tail naturally falls have little or no hair, and therefore need its protection.

'What does docking mean? It is taking off several joints of the tail, and then *searing the bleeding stumps with a hot iron.* Can any man imagine this torture applied to his finger, rendering it for ever less useful than Nature intended, and deliberately approve of its being inflicted on a helpless animal to gratify a passing fashion?

'I am informed that at Carmarthen a conviction

was recently obtained and a fine inflicted for docking a tail under circumstances of great cruelty, the operation taking nearly half an hour in all, and the irons having to be heated three times before the bleeding could be stanched. Was not that revolting barbarity? and who is to say that such cases do not frequently occur, and that in a greater or less degree they must occur so long as the custom prevails?

'I am informed—all credit to the humane person who so ordered—that by a recent regulation remounts for the cavalry are rejected if thus maimed. Might not this principle be extended, and all docking (except in case of disease or deformity) be considered to come under the term of cruelty in a legal sense, and therefore punishable by law? I am anxious to do what is right, and remain, Sir, your obedient servant,

'A Puzzled J.P.

'February 2, 1884.'

The regulation to which the puzzled J.P. refers has recently been repeated, September 12, 1884, and is as follows:—

'DOCKING HORSES.

'As it has been brought to notice that in the mounted service the regulations in regard to the length of horses' tails are not observed, and as horses with short tails are practically unfit for service in the field where flies are troublesome, the following addition has been made to Par. 5, Sect. 11 :—" Queen's Regulations and Orders of the Army. General officers, when making their inspection of mounted corps, will ascertain whether these instructions are strictly followed, and they will specially report every instance in which they are disregarded that comes under their notice. Horses with short docks are not to be purchased as remounts."'

The very forcible statements of the 'Puzzled J.P.' were supplemented by a letter from Mrs. H. McIlquham, of Staverton House, near Cheltenham, dated February 4, 1884 :—

'I am sorry to be able to add to the horrors contained in a "Puzzled J.P.'s" letter by saying that the terrible operation of docking the tails of horses is sometimes followed by lock-jaw. I was recently told of three such cases occurring in two stables.

'When lock-jaw occurs, it is usual to sling the poor animals, to prevent them from ending their sufferings by dashing their heads against their mangers. This wretched ending occurred in one of the three cases I have spoken of. The whinnying of the poor creature was touching to hear, and its sufferings will prevent a repetition of a similar barbarity in its owner's stables.'

It is really wonderful how the ever-changing fashion of the day perverts the judgment and blunts the feelings.

There is now before me a letter from a gentleman who is a great breeder of hunters. He acknowledges that he has himself lost two horses from lock-jaw produced by docking, and that he knows of other cases. Yet he declares in favour of the operation, because in his opinion 'it improves the horses very much!'

Again I say, as I did with regard to the gag bearing-rein, that even if the operation of docking *did* improve the appearance of the horse instead of disfiguring it, as is really the case, we have no right to inflict such torture on an animal merely to gratify our eyes.

There is even less excuse for docking than for the bearing-rein, for at all events the advocates of the latter do really think that it holds up a stumbling horse and prevents him from running away, besides 'improving' his appearance, whereas no such argument can be urged in favour of docking.

The Royal Society for the Prevention of Cruelty to Animals ought to be empowered to proceed against any one who was guilty of this abominable practice, and ought to carry out the law in the most rigid form, without the least respect of persons.

If the managing body of the Royal Society for the Prevention of Cruelty to Animals should be afraid to proceed in such cases because some of the delinquents belong to the Society, and their subscriptions would be lost, I can only say that the Society would not be doing the work which it professes to do. It ought to proceed against Lord Kennaquhair for having the tails of his horses docked, or for using the gag-rein, as fearlessly as against the butcher's lad for beating his horse about the head with his meat-tray.

Are we to revert to the practice of 'nicking' the tail? Perhaps the reader may not know what 'nicking' signifies. Indeed, I find that very few people have the least idea that 'docking' means anything more than cutting the hair too closely, and can scarcely believe me when I tell them that the very specious word 'docking' signifies the amputation of several vertebræ of the tail, and the searing of the raw and bleeding stump with red-hot irons.

There is only one abomination which has not as yet revived, and that is, the operation called 'nicking,' which was in practice some sixty years ago.

As might be expected, the demands of the professional eye became more and more exacting. Even the docked tail was not sufficiently distorted from Nature's model to satisfy that eye, which required that the tail should as far as possible resemble a half-opened fan in shape, and that instead of hanging down, as Nature intended, it should be stuck up perpendicularly.

So, after several vertebræ had been cut off, and the stump seared according to custom, the remaining vertebræ were cut partially through on the under

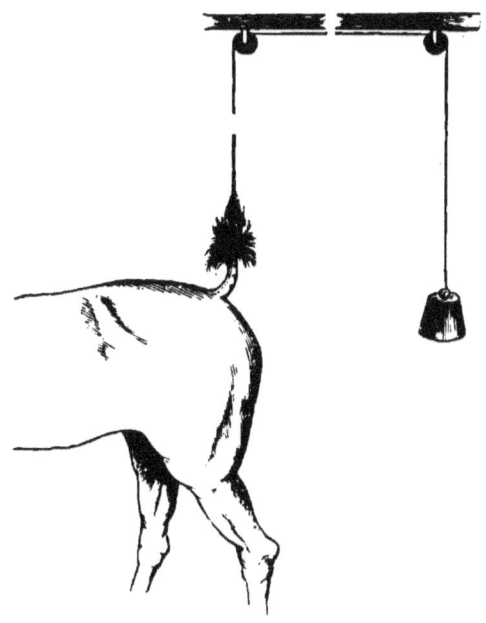

'NICKING' THE TAIL.

side. This was for the purpose of severing the rather powerful tendon which pulls the tail down. Lest the wounds should be healed in the ordinary manner, and so give the animal a chance of lowering its tail, a piece of machinery was invented and used as seen in the illustration.

The tail being tied up as here shown, it remained nearly perpendicular whether the horse stood or lay down.

Any of my readers who have ever suffered from a cut finger may form some idea of the agony which the horse must have endured while these wounds were being healed. The natural tendency of a wound is to contract in course of healing, and even during that natural process a considerable amount of pain is suffered. What it must be when the mouths of a series of wounds are kept mechanically torn apart we can scarcely imagine.

Some years ago, in the course of a lecture delivered at Cheltenham, I described this process, and illustrated the description by sketches on my black canvas. There were some very decided expressions of natural incredulity among the audience, several persons being unable to believe that such cruelty could have existed, and thinking that I had been imposed upon.

After the lecture was finished, a venerable, white-haired and white bearded gentleman came on the platform and asked leave to say a few words. He was evidently well known, as his appearance was greeted with loud applause. He said that he had heard some persons expressing disbelief, but stated that when he was a boy he had seen the operation of 'nicking' performed, and that both the description and sketches were perfectly correct.

I then found that he was Mr. Samuel Bowley, so well known as an advocate of teetotalism.

The illustration is taken from a plate in an old work on horse management. In the plate several horses are represented, so as to show the different stages of the process.

The reader could hardly imagine that anyone at the present day could have the hardihood to give any reason, except the following of fashion, for docking horses. Yet, in July of this present year (1884), a council of veterinary surgeons voted unanimously that *the docking of horses' tails was conducive to human safety*, and therefore ought to be adopted!

'Conducive to human safety,' forsooth!

What connection can there be between human safety and the length of a horse's tail?

Certainly the connection is not very easy to trace, and the train of reasoning is rather circuitous. However, here it is.

Supposing that you were driving a long-tailed horse, and that you were a careless handler of the reins, and let them hang too low, the horse might whisk its tail over the rein so as to take it out of your command, and then it might be frightened, and then might run away, and then might upset the carriage, and then you might be killed.

Even if all this long series of misfortunes did occur, the fault would surely lie with the careless driver. The 'Puzzled J.P.' is perfectly right when he says that carriage accidents were not more frequent when the horses wore their tails of their natural length than is the case at the present time.

Another 'reason' is, that the full and flowing tail is not easily put through the crupper. In fact, to the professional eye the tail is nothing but a convenient peg on which the crupper may be hung. So, to carry out these ideas logically, the rat-tailed horse is your only animal. There is no troublesome hair to get in the way, and by cutting off the vertebræ to suit your individual taste, you can make the crupper peg of any length that you please. Were it not for the crupper, the simplest plan would be to cut off the tail altogether, but that appendage was so evidently created for a crupper-peg, that enough of it must be left to serve its obvious purpose.

I very much wonder that the nicked and stuck-up tail was not voted to be 'conducive to human safety.' The process of reasoning would be very similar, and quite as sound.

A bad rider might mount a horse with a natural tail, and the horse might rear, and the rider would slip off over the tail and fall, and might be trodden upon and killed. But if the tail were nicked and stuck up, the rider would be caught upon it, and then he could hold on by the saddle as long as the horse reared, and regain his seat when the animal came on all fours again.

HAVING caused as much harm as possible to the foot and the mouth, and the eye and the neck, and the spine and the tail, the professional eye casts its baleful glance upon the Body, and demands that it shall be deprived of its natural covering just when it most needs protection.

The horse, like many other animals, has two sets of clothes: a cool and light suit for the summer, and a warm and heavy coat for the winter.

In the course of nature these clothes are regularly exchanged, and the horse puts on his long-haired winter dress before the cold weather comes on. But, unfortunately for the horse, the long, natural winter coat is to the professional eye as great an abomination as the long natural mane and tail, and so, either by clipping or singeing, the warm winter coat is cut as short as that which is intended only for summer wear.

Here, then, is another of the many ingenious devices by which man does his best to shorten the life of the horse. The animal, whose skin is singularly sensitive, and given to perspiration, is taken on a cold winter's day out of a warm stable, in which all the pores are kept open, and brought into a frosty temperature, which is often aggravated by a sharp wind. Only one result is likely to take place, and mostly does take place. The perspiration is checked, thrown back upon the system, and the horse takes a cold, which, unless promptly treated, will develop into pneumonia, or inflammation of the lungs.

In America, horse owners are wiser than we are. They have a complete set of waterproof clothing for each horse, even covering the ears, and these coverings are made in several portions, are ventilated, and shelter the horse completely from the cold air. There are apertures through which the shafts and reins pass, but even these are guarded with waterproof flaps.

In this country, however, we are less thoughtful, and many a fine horse is lost through the want of a little forethought.

As we have already seen, 'reasons' are plentiful enough when any outrage on Nature is committed, no matter whether the subject be ourselves, or the animal of which we have the control. All kinds of 'reasons' have been proffered when man acts in contravention of Nature.

Shaving, for example, has its advocates; its

object being to make the masculine face as effeminate as possible. Our immediate ancestors, say of the reign of George II., were more logical than ourselves, and not only shaved the face, but the entire head, substituting an artificial covering of horsehair or goat's hair, called a peruke, for the natural covering which they removed.

The ancient Egyptians were still more logical, for they shaved the whole body daily, from the crown of the head to the sole of the foot. In like spirit we cut away the natural protection of the horse's hoof, and substitute an artificial protection of iron.

Many of my readers can remember the time when the British soldier was obliged to buckle round his neck a stiff leathern stock, so that he could hardly breathe, and men were constantly struck down in the ranks with incipient apoplexy. No sooner did common sense urge the abolition of the stock than the old officers arose in their wrath, and eulogised the leather stock to the skies.

It saved the soldier from a sword-stroke should he happen to be in personal conflict with a dragoon, and the dragoon should happen to strike his neck, and the blow should happen not to be parried. It saved the soldier the trouble of holding his head up, because the stock did it for him. In point of fact, the professional eye was accustomed to the stock, and could not recognise a man as a soldier unless he wore it.

As for feminine humanity, words fail to express the countless methods by which women have done their best to cripple themselves for life, and the extraordinary ingenuity of the excuses—'reasons' they term them—with which they defend tight corsets. I have seen one which was made entirely of steel, like a cuirass, and which opened at one side with hinges. Tight boots, with high pegs under the centre of the foot instead of heels; hoops in the days of George II., crinolines in the time of Victoria, poisonous metallic paints covering the face, neck, and arms, and a hundred other absurdities, all had their advocates and their 'reasons.'

Among the reasons which the groom urges in favour of clipping, and which his master accepts as true, is that it is impossible to keep his horse dry if it be allowed to retain its winter coat, and that it is almost impossible to clean the horse from mud after he has been out on a wet day.

Here unwittingly the groom betrays the real reason of his advocacy of clipping. He is saved trouble by it, no one denying that it is easier to clean a short-haired than a long-haired animal.

The wetness of which the groom complains is simply the result of imperfect ventilation in the stable. Now, the uneducated classes are, as a rule, absolutely ignorant of the value of ventilation in their own houses, and therefore can hardly be expected to trouble themselves about a constant supply of fresh air in the stable.

In order to obtain enough air for its enormous lungs, the horse is obliged to breathe rapidly. Then the skin comes to the rescue of the lungs, the functions of the skin and lungs being co-ordinate, and a copious perspiration is the result, no matter what the length of the hair may be. But, in the morning, the groom finds that to clean and dry the long-haired horse is a tedious business, and so he induces his employer to have the horse clipped or singed. So universal is this custom that as a rule the owners of horses would as soon have their hoofs unshod as their coats unclipped.

Another 'reason' which grooms advance for clipping is, that the horse moves much more freely after than before the operation, because it is relieved of the weight of superfluous hair. Mayhew (p. 235), with all the caution of a man of science, does not actually deny that such an effect may be produced by clipping, but he very emphatically says that he never saw it.

Indeed, if the hair which is removed by clipping were put into the scales, its weight would prove to be very trifling, and certainly insufficient to make any perceptible difference in case of movement. Moreover, the hair so removed was distributed over the whole body and limbs, and not hung in any spot where it might cause hindrance to action.

Yet another 'reason' is advanced by grooms. 'We are obliged to have our hair cut regularly, and so ought the horse.' But the hair of man, like that of the mane and tail, is permanent, and not deciduous, and so the argument is beside the point at issue. If we were gifted with summer and winter heads of hair, there might be some force in it, but, as the reader will easily perceive, the two cases are not parallel. Moreover, women, as a rule, wear their hair long.

Mayhew concludes his long and elaborate indictment against the practice with the following words of warning :—

'The advent of the summer coat is delayed, and the system seems to suffer greatly during the subsequent period of changing the coat. The pace flags, the spirits fail, and the quadruped becomes more susceptible to disease at a time of year when equine diseases are commonly more general and more virulent.

'The master who makes the welfare of his horse subservient to the idle prejudices of his groom, is fitly punished in the lengthened period of his animal's compulsory idleness, appropriately finished by the payment of a long bill to the veterinary surgeon.'

Here I must call attention to another point. I have already mentioned the cautious manner in which Mayhew, as a man of science, writes when he treats of a subject which is outside his peculiar province. This very caution gives additional weight to the absolute decision with which he writes when on his own ground. As to the question about freedom of movement being the result of clipping, no matter how absurd such a statement may seem, he writes as follows :—' This may be the fact, although the author has hitherto seen no such marked change follow the operation as will allow him to deliberately corroborate the general assertion.'

But when he treats of the evils which clipping inflicts upon the horse, he uses no such caution, but employs the most direct and positive language :—

'Horses which have been clipped or singed *are* (not may be) rendered more susceptible to many terrible disorders. Any internal organ may be acutely attacked, because the perspiration has by exposure of the skin been thrown back upon the system. Numerous hunters (which animals are always clipped) *fail* (not may fail) at the beginning of the season from this cause.'

This is the voice of one who speaks as having authority.

Clipping is quite a modern invention, and was introduced into this country from the Continent about 1825, our officers having become acquainted with it during the Peninsular War.

It has been said by some writers on the subject, that although they are entirely opposed to clipping, and would never allow a horse to be clipped if they had been in charge of it from the first, yet, if a horse be once clipped, it is impossible to discontinue the practice, as the coat will become rough and staring.

I confess that I cannot agree with this axiom, nor understand the theory on which it is based. The coat of the horse is changed in accordance with the season of the year, and no matter what may be the ill-treatment of the coat of one year, it cannot affect that of the next year.

Some thirty years ago, when common sense began to prevail over conventionality, and men began to ask themselves whether their Creator made their beards for the purpose of having them cut off daily, a similar reason was urged against the abandonment of the razor. It was all very well for youths who never had shaved, as their moustaches and beards would possess a most becoming gloss and silkiness. But, when once the razor had cut off the original hairs, the after growth would be coarse and stubbly.

This was a very plausible argument, and, until

examined, seemed to be carried out by facts. It was true that the first growth of a beard is soft and silky. It was equally true that when a man's beard is of two or three days' growth after shaving, it is coarse and harsh, and if the beard be a dark and stiff one, looks very much like a ragged blacking brush. But, when the beard has grown to the length which it would have attained if its wearer had never shaved, it will be just as soft and silky as if it 'never had known the barber's shears.'

If this be the case with the permanent hairs of the human beard, which are analogous to those of the horse's mane, much more so must it hold good with the deciduous hairs of the horse's body. We find exactly the same in birds, the perfection of whose plumage after moulting is not in the least marred by injuries done to the feathers before the moult.

Boys who for the first time own a magpie or a jackdaw, generally discover this fact by experience. As soon as they obtain possession of their bird they clip its wings, and are under the impression that they have prevented it from flying during the remainder of its life. But, unless the boy has managed, as he ought to do, to attach his bird to himself so thoroughly that it does not wish to leave him, he will find that after the moult is complete, the bird regains the whole of his plumage, and can fly as well as ever.

A younger brother of mine fell a victim to his ignorance of this part of natural history.

When a boy, he became the happy possessor of a jackdaw, and, having heard that the wings of birds ought to be clipped, he proceeded to clip them, carrying out the work most conscientiously. He cut off the feathers of both wings close to the limb, and the tail close to the body, beside snipping away here and there feathers which looked too obtrusive, and was as satisfied with his work as is a farrier after scooping out a hoof.

Soon afterwards came the moulting season, when, as a matter of course, the bird regained its plumage, and flew away.

All the cutting and clipping to which the feathers had been subjected had no effect upon the next suit of feathers, and neither does clipping have any effect on the next set of hairs. I lay some stress on this fact because many persons have been dissuaded from the abandonment of clipping because they have been told that if the practice be once began, it cannot be relinquished without making the coat rough and staring. Similarly, many persons who have been accustomed to shave are afraid to abandon the razor, because they have been told that their beards will be coarse, dry, and rough.

## CHAPTER XVI.

The lungs of the horse and their comparative size—Their shape and position—Their demand for air—Defective stables—Ventilation—A shining coat—The stomach of the horse and its small size—Comparison with the stomach of the ox—Mistakes in feeding—Result of overloading the stomach—Experience of a veterinary surgeon—Water, when to give, and how much—Traditions of trainers—A Turkish custom—Purity of water and water vessels—Sloping floors and their evils—The locomotive and the horse again—A sloping couch—The manger and drinking trough—Structure of the head and throat—The stable door—Width of stalls—'Weaving' and crib-biting—The electric manger.

IN the last chapter I had occasion to mention the ill-ventilation which is the rule in most stables. This brings us to another and most important point in the management of the horse. In this work I have been careful not to trouble the reader with needless anatomical details, and have only given those which bear directly upon the relationship between man and the horse. Putting aside the heart, which is less liable than any of the internal organs to be injured by man's mismanagement, we find two internal organs of equal importance—namely, the lungs and the stomach. Supposing we were to split a horse longitudinally, there are very few persons who would not be surprised at the enormous dimensions of the lungs and the very small size of the stomach.

Supposing a water-melon to represent the lungs, an orange would represent the stomach. Or supposing that we take the word 'Do,' the capital letter would in size and shape represent the lungs, and the small 'o' would equally correspond to the stomach. Then it must be remembered that the lungs are double organs, while the stomach is single, so that the discrepancy in their comparative sizes is enhanced.

Both these organs are mismanaged by man, but in opposite ways, one having its necessary supplies cut short, and the other being supplied with more nourishment than it can assimilate at the time.

First we will take the Lungs.

If we examine the size of the lungs when compared with the body of any mammal, we shall find that the larger the lung, the higher is the vitality, and, in consequence, the more air has to be consumed.

Now, in the horse, the lungs are simply enormous. Their outline very much resembles that of the capital letter D, except that the upper corner requires to be lengthened upwards to the right. Perhaps a better idea of the outline may be obtained from the national harp as shown on the Irish shield of the present florin. This harp must be large enough to reach beyond the withers backwards, then extend diagonally to the elbow, and then upwards to the junction of the neck with the breast.

Let the reader turn to the skeleton on page 3, and then draw a line from the top of the rib just below c, and continue it diagonally to I. This line will represent the 'sound-board' of the harp, the 'pillar' occupying the front of the chest, and the 'neck' running under the spine. Nearly the whole of the cavity of the ribs in front of the 'sound-board' is filled with the lungs.

So enormous a structure indicates that the supply of fresh air ought to be correspondingly great, and that there shall be means of escape for the air which has been breathed, and in consequence is not only useless for respiration but absolutely poisonous.

Yet, in how few stables do we find even an approach to systematic ventilation? The reason is evident enough. In the first place, a stable is considered so simple an edifice that any one can build it. Accordingly we find that, as a rule, the architecture of a stable is about on a par with that of a house which a child builds with his box of wooden bricks.

The typical stable is an oblong box of bricks, divided into two storeys, the upper being used as a hay-loft, and connected with the stable by a ladder and a trap-door. At the end is the door, and on one side is a little window. Opposite the door is the manger, and above the manger is an open rack, into which hay can be pushed from above.

The floor slopes from the manger so as to carry off moisture into a gutter which runs at right angles with it, and then the builder thinks that he has produced everything that a horse ought to require. In point of fact, however, if he had deliberately set himself to undermine the horse's health, he could not have been more successful.

But no one expects that the builder of a stable is likely to know anything of the structure of a horse, or, if he did, to depart from the rules of custom.

The builder has no idea of the true functions of respiration, or of the poisonous character of air that has been once breathed. So he makes no provision for the admission of fresh atmosphere nor for the escape of foul air. He neither knows nor cares that the air which has been breathed, additionally laden with the pungent ammoniacal vapour that fills an unventilated stable, passes through the hay in the rack and thence into the hay in the loft, infecting them both.

It is nothing to him that lungs were not intended to breathe effete air or the vapour of ammonia. It is nothing to him that the small size of the windows keeps the horse in semi-darkness, so that he is half blinded when he goes into it on a bright day, and quite dazzled when he comes out. It is nothing to him that the sloping floor is a perpetual strain upon the muscles of the legs during the time when they ought to be at rest. He has built his stable as stables always have been built, and the creature which is to inhabit it has nothing to do with him.

One of the chief difficulties with which an owner of horses has to contend when he desires ventilation is, that the groom, as a rule, dislikes ventilation, and cannot distinguish between fresh air and a draught. One of the evils of imperfect ventilation has already been mentioned in connection with the custom of clipping.

The groom will tell you that a ' warm '—*i.e.* a close —stable makes a horse's coat shine like satin. So it does; but as it mostly kills the horse, the benefit is rather doubtful, and reminds one of Charles Lamb's Chinaman, who burned down his house whenever he wanted to roast a pig.

It would be better for the horse to live, as the North American pony lives, in a far severer climate than ours, never to be housed and never to be groomed, than to pass more than half its time in such a pestilential atmosphere as that of an ordinary stable when managed by an ordinary groom.

After glancing at the lungs, we come to the Stomach.

In proportion to the size and weight of the animal, the dimensions of the stomach are wonderfully small, the combination of large lungs and small stomach showing that the animal is intended for speed as well as strength.

I am quite sure that not one groom in a hundred (or say a thousand) has the faintest idea of the difference in size and structure between the stomach of a horse and that of an ox, or that the mode of feeding is essentially different. Yet there are plenty of horses and oxen feeding in pasture lands which can be seen by any one who has the use of eyes.

The ox, having no teeth in the front of the upper jaw, simply squeezes a tuft of grass between the lower incisor teeth and the hard pad which takes their place in the upper jaw, and with a sort of jerk tears off the herbage—'*licks* up the grass,' according to the Scriptural phrase. The tuft of grass thus torn away is roughly bitten into a sort of ball, and passed into the curious set of chambers into which the stomach is divided. When the animal has obtained a sufficient supply of grass for a meal, it lies down, and returns the food by degrees into the mouth, and then masticates it.

But the horse, with his single and small stomach,

feeds on a different plan. It is always on the move, selecting and nipping the grass blades as it walks along, and masticating the food before it swallows it. Thus, the ox feeds while still and recumbent, while the horse always eats while moving, and never while recumbent.

It is evident, therefore, that the horse ought to be fed frequently, and not in very great quantities at each meal. The groom, not being aware of this fact, is too apt to cut off the supply of food before a journey, in order to make the horse 'travel light.' Then, after the journey, wishing to be kind to the horse, he gives it a double feed by way of reward.

The natural consequence is, that the animal, ravenous with hunger, eats too rapidly, overloads its stomach, and suffers accordingly. Even human beings, in spite of knowing better, will do the same unless restrained by others, and we cannot expect a horse to be wiser than a man. Cases are not unknown where the coats of the stomach have given way, and the horse has died in consequence of the injury.

Sometimes, especially in agricultural districts, a peculiar disease is produced by overloading the stomach, and generally makes its appearance on Monday morning.

Both the horse and the labourer who is in charge

of it are at work in the field for twelve or fifteen hours daily throughout the week. The man is naturally so wearied with the day's toil, that after the work is over he is fit for nothing but to take his supper and go to bed, where he instantly falls asleep, and does not awake until it is time for his next day's labour to commence. Even on Sundays he has to look after his horses, so that he does not even get the one day's rest in seven to which he is entitled.

So it often happens that the man, not knowing the distinction between the structure of the horse and that of himself, supplies the animal on Saturday night with enough food and water to last until Monday morning, and then has the Sunday to himself. On the Monday, when he goes to the stable, he finds the horse with one of its legs swollen beyond all compass, and held off the ground; all the water gone, and the animal half mad with thirst.

Even with the most careful attendance, many days must pass before the horse is again fit for work. We can hardly blame the man, who treats the horse as he himself would like to be treated—*i.e.* by being supplied with plenty to eat and drink, and left to undisturbed repose for the whole day.

The man has acted in sheer ignorance, and so far is free from blame. But why should he be ignorant of so simple a matter? The real fault lies with the owner, who has entrusted his horse to a man without having ascertained that his subordinate knows anything about the animal.

So universal is this custom of keeping the horse too long without food and then trying to make up by over-feeding it, that even in high-class stables there are few horses which have not more or less suffered from it. A veterinary surgeon of long experience stated that he 'never dissected the carcase of an aged animal without finding the capacity of the stomach morbidly enlarged, and the walls of the viscus rendered dangerously thin by repeated distension.'

A groom who knows anything of the structure of the horse will always bear in mind the all-important maxim in feeding the horse—*i.e.* LITTLE AND OFTEN.

Then there comes the question of the amount of WATER which a horse ought to have, and when it ought to have it. The answer is simple enough. Let the animal drink when it likes and as much as it likes, and it will never damage its stomach by over-drinking.

In this country there still prevails an idea, or rather a superstition, that if a horse be allowed to drink freely before starting on a journey, he will become broken-winded by the end of it. I say especially 'in this country,' because in America no such prejudice exists. As might naturally be supposed, the deprivation of water conduces to the very evil which it was intended to prevent. When the thirsty animal does get at the water, it drinks eagerly and rapidly, just as a thirsty man would do, and so causes the stomach to be abnormally distended.

The Americans seem to have taken a leaf out of the Turk's book, as far as the giving of water is concerned.

'While on the subject of horses, let me say a few words upon their management by the Turks.

'First of all, a Turk never is wantonly cruel to any beast. He never strikes nor spurs a horse in a rage, and his movements are so quiet and collected, that a horse soon forgets all fear with him, and to find a restive horse is rare indeed.

'Whenever a Turk passes water, winter or summer, he lets his horse drink as much as it wants; but when he has done so, he rides it on fast at once, and never lets it stand still after drinking. If possible, he gives it its fill of water half an hour before the end of its journey. Directly he dismounts, he loosens the girths, and then leads the horse about till it is quite cool. It is then put in the stable with the saddle on, and this is not taken off for an hour or more. When this is done, and when he has had water on his way home, he never "breaks out," and never refuses his feed.

'It is a British prejudice, and a very cruel one, not to let a horse drink when he is thirsty. If any of my readers have a horse that does not feed when he comes in from a day's hunting, let him try the Turkish plan.'—H. C. Barkley, 'Five Years in Bulgaria.'

Not so many years ago the same idea prevailed in connection with the training of competitors in athletic sports.

As some of my readers may have personal reasons for remembering if they have rowed in a college boat-race, one of the chief duties of the trainer was to dole out liquids to his men as if they were adrift on the ocean with a scanty supply of water. The men were kept in a perpetual state of raging thirst, so that on the day when the traditional black draught had to be administered, they almost flew at it because it was liquid.

At the present time, the trainers, being educated men, have seen the absurdity of a process which was invented by the ignorant prize-fighting set of a century ago, and have gradually emancipated their charges from the incessant physicings, 'training mixtures,' semi-raw beef, total deprivation of vegetables, and the chary doles of 'old,' *i.e.* hard ale, which constituted the whole of their liquid nourishment. No wonder that so many broke down under the process of 'going through the mill' as it was called, and that their constitutions were permanently injured.

But, when horses are concerned, the training is very little improved, the trainers, as a rule, being illiterate men, and therefore clinging to the traditions of their forefathers. Grooms and others who are in charge of horses naturally look up to the managers of racing stables as superior beings, and accept unconditionally any of their dicta as necessarily infallible.

It is time, however, that we should fling aside all those traditions, and that in England, as in America, a horse shall no longer be tortured with thirst during his work.

In many large establishments the horses are allowed to drink whenever they like and as much as they like. One notable example is to be found in the stables of Messrs. Barclay and Perkins.

When a horse wishes to drink, he rattles his halter in a peculiar manner which the horses learn from each other. The groom in attendance knows the signal, and lets the horse loose. The animal goes of his own accord to the water-troughs, drinks as much as he wants, and then returns to his stable, without requiring an attendant.

## PURE WATER. 299

This principle is, I believe, adopted in many places, but I mention the establishment of Messrs. Barclay and Perkins because I have been eye-witness of its practical working.

Above all, the water and the vessel in which it is contained must be scrupulously clean. Too much attention cannot be given to this injunction, especially in stables where there is no water trough, and the pail is the only means of giving the horse drink. Those who are much about stables will understand what I mean. No nitre or other farrier's messes should be allowed in the water.

I very much wish that a few boldly drawn coloured diagrams could be issued, showing the principal points in the structure of the horse's foot, mouth, lungs, and stomach; that it should be incumbent for these diagrams to be kept displayed in every stable, and that no one, not even a boy, should be allowed the management of a horse until he could show that he was familiar with, and could understand, the diagrams.

AMONG the many faults of the ordinary stable, I casually mentioned the sloping floor. Of course, the object of this slope is evident enough, and it is also evident that if the floor were perfectly level it would be impossible to keep the stable decently clean. But the slope is in the wrong direction, and its effects upon the horse are simply disastrous.

As long as the horse stands or lies upon a floor that slopes backwards, he can enjoy no real rest. Suppose that we revert to our original parallel between a horse and a locomotive engine, and suppose the engine to be placed on a rising incline. The result would be that the engine would immediately begin to roll backwards; and if it were to retain its position, it must keep the wheels at work so as to counteract the incline. I do not think that we should much admire the wisdom of the owner of the engine if he were to station it all night upon an incline, so as to keep it at work, and wear out its machinery merely in order to retain its position.

Yet, in most stables the horse is subjected to precisely similar treatment. By the force of gravity it is kept perpetually slipping backwards, and can only counteract this slipping process by the continual action of the muscles.

Let any one try it for himself. Let him stand for a quarter of an hour upon the side of a hill and facing the slope. He will very soon find out that, however still he may stand, he cannot retain his position without exerting the muscles of the legs to a degree which every minute renders more irksome. Now let him turn in the opposite direction, and he will find that the relief will be instantaneous, and that he need not use one-tenth of the exertion which is required in the former position.

Instinctively the horse knows this fact as well as man does by means of reason, and when the animal is at liberty on sloping ground, the animal invariably stands so that the fore-feet are lower than the hinder pair.

Even when the horse lies down for its night's rest, matters are not mended. Let the reader try to sleep on a sloping bed, and he may have some idea of the feelings of a horse under similar circumstances.

Of course there must be drainage, and the drains must slope downwards. But there is not the least necessity that though the drains slope, the floor should slope also; and any one who possesses the least smattering of sanitary engineering can drain a stable more effectually than can be done by the present system, and yet can keep the floor level, so as to make it a comfortable resting place for the horse, whether the animal be standing or lying.

THERE is one point more on which I should like to touch, though it is not nearly so important as those which have been already mentioned. This is the position of the manger and drinking trough. For many years I have felt certain that both these accessories to the stable are placed much too high for the convenience of the horse, but should scarcely have felt courageous enough to publish my idea, had not several acknowledged 'horsey' authorities expressed the same opinion.

Watch a horse feeding and drinking when he is at liberty, and you will see that when he feeds his nose is on a level with his fore-hoofs, and that when he drinks it is below them. The whole structure of his head, neck, throat, and especially that of the veins of the neck, indicates the position which the head was intended to assume when the horse was eating or drinking. Yet we put the hay into a rack above the level of his head, throw the oats, beans, &c., into a manger on a level with his breast, and pour the water into a trough on the same level as the manger. In the model stable, the manger would be on a level with the floor, and the running water in a trough a little below it.

ONE more defect of ordinary stables has yet to be mentioned. The door is, as a rule, much too small, both in width and height, especially the former. This is not of so great importance when the horse is entering the stable, as when it is coming out. A high-spirited animal, which has passed some eight hours in the stable, is full of spring and joyfulness when brought out into the air, and dances, rather than walks, through the door. If the doorway be too narrow or too low, the animal is apt to strike itself against the door-post, and so to break one of the projecting portions of the hip bone. It is true that if a groom places himself in front of the horse and backs out of the stable, leading the horse after him, he will take the animal safely through a very narrow doorway. But it is never right to trust the discretion of a subordinate when the welfare of the horse is at stake, especially when a mere enlargement of the doorway will make it safe even for a restive horse and a careless groom.

There are many other defects in ordinary stables, such, for example, as the narrowness of the stalls, which is the primary cause of 'weaving'—*i.e.* fidgetting, crib-biting, and other 'vices' of the stable.

As for weaving, all imprisoned animals weave in one way or another, as any one may see by visiting the Zoological Gardens. It is their only mode of relieving the intolerable monotony of their existence. There is just now a wolf which has invented a totally new method of weaving. He goes to the back of the cage, runs a pace or two, jumps into the air with outspread and stiffened legs, and comes down so as to slide as far as the bars.

There is really no great harm in weaving, any

more than there is in a child swinging its legs if compelled to sit still on a form or chair when its natural instinct urges it to run, and jump, and shout, in order to let off its superabundant energy.

In the one case the groom, and in the other the teacher, looks upon these restless movements as a sort of wilful vice, which must be checked by punishment. Yet, the fault does not lie with the horse or child, but with those who deprived it of its natural amount of exercise.

As to crib-biting, a vice which is mostly thought to be incurable, it is almost wholly caused by the narrowness of the stall, and the confinement of the movements of the animal. I never heard of a horse beginning this vice in a loose box. It may from custom continue it, though the reason for it has been taken away; but, as far as I know, the evil habit has never originated in a stable where the horse could move about.

I have seen a very ingenious form of manger, which was invented in order to cure horses which were given to crib-biting. A galvanic battery was connected with metallic plates which ran above and below the edge of the manger. As long as the horse uses the manger for its legitimate purposes, he has no reason for supposing that it is not like any other manger, but, as soon as he seizes the edge with his teeth, he forms a connection between the upper and lower plates, and receives a shock which makes him loosen his hold. It is not strong enough to injure the animal, but quite strong enough to impress on his mind that mangers are not meant to be bitten, and that if he bites a manger it will avenge itself. This invention was shown me by Messrs. Martingale, of 158 Piccadilly.

## CHAPTER XVII.

The horse and the locomotive again—'Vice' in horses and its invariable cause—Mayhew's opinion—Vice in cavalry horses—The soldier and the 'irreclaimable' savage—New mode of treatment—Apparent failure and ultimate success—A relapse when in strange hands—Another 'irreclaimable' savage—Story of 'Fly,' 'The Baroness,' and 'War Eagle,' all three being New Zealand horses—The horse's capacity for affection—Its peculiar love for man—The horse a gregarious animal—Bulgarian horses—'Spoiled' horses—The horse's desire to obey man—A circus horse—Gilpin redivivus—Cavalry horses and their habits of obedience—The old horse at a review—Disbanded horses in a thunderstorm—The 14th Hussars at the Cape—Escape of their horses—An amateur review without officers—Muster of loose and wounded horses after battle—Mr. Luck's horse—Imprisoned in winter—Endurance of the horse—'Sam' and his tricks.

THROUGHOUT this work comparisons have been drawn between the horse and the locomotive engine. It is worthy of notice that whereas the driver of an engine comes to look upon it as a living creature, always talks of it as 'she,' and treats it as if it were possessed of feeling and intelligence, too many drivers or managers of horses look upon them as machines which can only be made to work by hard words and harder blows.

They think—or, rather, they assume, as the ignorant always do in lieu of thinking—that horses and men are naturally antagonistic to each other; that the horse is always trying to thwart the man, and that it will only work when compelled by the terror of the lash. They act on this assumption, and the natural consequence is, that there is a perpetual struggle between them and the horses which are under their control.

Of course there are exceptions, but, as every one will admit who has had experience of grooms, stablemen, and drivers, whether in country or in town, brutality is the rule, and kindness the exception. It must be said on behalf of the men that they have the excuse of ignorance. They have been brought up to the idea that a horse can only be ruled by fear. They have never seen any other mode adopted, and naturally refuse to believe that any other mode is possible.

I have often wondered why, even for their own sakes, those who have the charge of horses do not invariably treat them with kindness. It is so much easier to manage a horse that is loving, confiding, docile, and obedient, than one which goes in constant fear of its attendant, and has to be coerced into every action, that a groom who deliberately makes a horse afraid of him is infinitely more foolish than the poor beast which has no choice in the matter.

For my part, the longer I live the more fully I am convinced that 'vice' in a horse signifies cruelty on the part of man. Vice forms no part of a horse's normal constitution. No horse is born with a desire (like Mr. Winkle's natural taste for perjury) for kicking or biting human beings. An 'irreclaimable' savage is made, not born, and the most painful feature in the case is, that the fiercest savages are invariably high-spirited and sensitive animals.

All horses are not alike, as the ordinary groom thinks, any more than all boys are alike, as the ordinary schoolmaster thinks, or at least used to think. There is as much variety in the characters of horses as of boys, and the schoolmaster or the groom who tries to convert his charges into machines will always find himself in difficulties with the pick of the school or stable. It is the sensitiveness and high spirit of a horse that convert it into a savage, when treated with persistent brutal cruelty. The dull-witted horse submits and suffers, but the high-spirited animal rebels against such treatment, and soon learns that if it can do no more, it can at all events have the gratification of making its groom afraid of it. And, as the groom is the only human being with whom it is brought into contact, we need not be surprised if it is unable to distinguish his conduct as that of an individual, and in consequence considers that the whole of the human race is in league against it.

I used the term 'irreclaimable' as applied to a savage horse, but enclosed the epithet in inverted commas to show that it is a word to which exception can be made. I do not consider any horse, however savage, to be irreclaimable, but believe that if brutal and ignorant men have converted the horse into a savage, gentle and wise men can reclaim him from his savagery. Two more 'irreclaimable' savages never lived than the celebrated horses 'Stafford' and 'Cruiser,' the latter being about as safe to approach in the stable as a tiger in his den.

Lord Dorchester, the owner of 'Cruiser,' stated in a letter to the *Times* that until Rarey took the animal in hand he had not been ridden for three years, and that it was necessary to keep an iron muzzle always on his jaws. His paroxysms of rage would last for several days together, during which time no man dared approach him. He has been known out of sheer frenzy to kneel on the ground and tear up the road with his teeth.

Yet, it is a matter of history that both these animals were reclaimed in a very short time, and by the sheer power of kindness. I never saw 'Stafford,' but with 'Cruiser' I was on familiar terms, the beautiful animal being as gentle and playful as a kitten, and quite as craving for human notice. These qualities were not implanted by Rarey. They existed already, but had lain in abeyance until Rarey's gentle sway evoked them, and but for him would never have been suspected.

Mayhew and Lupton speak very strongly on this subject in their work on 'Horse Management:'—

'When a horse is tenderly and kindly looked after, it is difficult to form any conception of how interesting and intelligent he becomes, watching his groom's every movement in the stable day after day and week after week, until he knows as well as the man himself what is next to be done. How patiently such a horse will stand to be cleaned or saddled, and how eagerly will he turn round to have his head groomed or to be bitted, and how grateful is he but for a crust of bread, a handful of oats, a mouthful of hay, or even a kind word, when spoken in a kindly tone by his master.

'When such a man enters the stable the horse receives him with a neigh of welcome, pawing the ground with his forefoot until the master is alongside of him and pats him on the neck encouragingly.

'As good masters make good servants, so do good kind grooms make good and affectionate horses. I do not mean the man who makes the horse's coat shine like satin, for drugs will give him a glossy coat while they undermine his constitution. I refer to the groom who, in addition to the lustrous look of the animals in his care, is never heard yelling at them to " stand over," or " come round," who is assiduous in his attentions upon them ; who, recognising in them the possession of virtues and passions, and but scarcely less reasoning powers than he possesses himself, will never even commit an outrage upon their feelings, far less deny them merciful treatment.'

I have received many accounts of cavalry horses, showing the advantages to horse and rider of kindness on the part of the latter. If any man could have an excuse for being impatient with a horse, it is the cavalry soldier. He cannot spend as much time on his horse as he likes or when he likes. His time is meted out to him like his rations, and if within the given time he cannot produce his horse in a fit state to pass his superior officer's eye, he has to suffer for it.

So an ignorant man, who has always been used to seeing horses ruled by force, may be pardoned for carrying out his work in the only manner which he knows. But he has imposed upon himself a task which is far more severe than would have fallen to him had he ruled his horse by kindness. Here is one of the many narratives that have been sent to me, the writer in this case being the soldier himself. As, however, he was unused to composition, he was needlessly diffuse and encumbered with moral reflections, as is the way with inexperienced writers. I have therefore condensed his account and narrated it in the third instead of the first person. I may mention that the writer gives the names of the lieutenant-colonel in command, his major, his captain, and his adjutant, as references if they should be required.

He enlisted as a mere lad in a dragoon regiment, and after some three years of service found that his horse was not spirited enough for him. There happened to be in the same stable a very splendid horse, which was unfortunately so ferocious that the men were afraid of it as an irreclaimable savage and hated it.

The soldier had noticed that his comrades were apt to avenge themselves on the horse when they could do so with safety, and thought that a different treatment might improve its temper. So he asked the troop sergeant-major to let him have a trial, and, if he succeeded, to exchange the horse for his own.

The sergeant-major very properly remonstrated with him, saying that the horse had injured several men severely, and that his life would be endangered if he meddled with it. He also added the curiously grim military aphorism, that soldiers are wanted for service and not as corpses or patients in hospital. However, he at last reluctantly gave his consent.

The man at once went to take possession of his new charge, and almost immediately verified the sergeant-major's predictions. He had hardly begun to groom the horse, when it seized him by the side, lifted him from the ground, and banged him against the manger. He was rescued by his comrades, and though much bruised, was not seriously hurt.

The other men wanted him to beat the horse severely as a punishment for its conduct, but he steadily refused, and tried to conciliate the animal by kindness. It was some time before the horse could be made to understand that his new master did not intend to hurt him; but in the course of a month or two the animal's feelings underwent an entire revulsion, and it conceived the strongest affection for its master, following him like a dog, and allowing him to do anything with it.

After a while, the soldier had a furlough of six weeks, and went away, begging his comrades to treat the horse kindly, feed it well, and not trouble themselves about dressing it, an operation which it had always resented.

He had only been away a day or two when he received a letter saying his horse was in hospital.

On his return, he asked to see the animal, which had evidently returned to its vicious ways, and was challenged to enter its box without speaking. He unhesitatingly did so, and was cordially greeted by the horse, who knew his step.

He was allowed to take the horse to the troop stables, and found it in reality so well and strong, that two days afterwards it carried him through one of the late Lord Cardigan's very trying field days with even more than ordinary vigour.

It was found afterwards that the horse, as is usually the case, had been made vicious by brutal treatment. He was a well-bred, high-spirited animal, and had been placed in the regiment when very young. He was fortunate enough at first to fall into the hands of a kind master who made a pet of him, and in consequence he was docile and affectionate. But his master died, and he fell into the hands of men who looked upon a horse merely as a machine, and could not understand that it possessed feelings.

Not having been accustomed to ill-usage, the horse naturally resented it, and before long had learned to look upon man as its natural enemy, which could only be kept from hurting it by being made afraid of it. But, when at last a man refused to avenge himself as he might have done, and resumed the kind treatment to which it had been accustomed, its better nature revived, and it became docile and affectionate to its friend, though it still remained suspicious of others.

Another history of a savage horse I give in the writer's own language :—

'St. Kilda: Melbourne.

'While in New Zealand, I purchased from Australia a thoroughbred filly, "Fly." On attempting to mount her, I saw from the set of her back, the white of her eye, and the twist of her neck that she was a dangerous animal, and, being at the time delicate and nervous, I sent her to Mr. R——'s racing stables, to be ridden by one of his jockeys.

'She threw her first rider, C——, three times in one day. He had known her in Australia, and remarked that he would as soon see the devil come into his stables as "Fly." In a fortnight she was returned to me with a note to say that she would never be fit for any but a bold, strong rider, and that she must never have much rest.

'Still, in every other respect she was so much to my taste, that I did not like parting with her without another trial. So I fondled her, talked to her, got half on and then down several times, so that she might not be afraid of me; and when I did at last mount, I sat quite still until she was inclined to move.

'From that day no quieter animal could exist

*with me.* Unsaddled, and with only a halter instead of a bit, she would carry me anywhere. Once, a child ran under her feet. The thing was so sudden that I was powerless, but she at once stopped, and held up her hind leg so as not to hurt the child. Yet, when I was obliged to sell her no one could ride or do anything with her.

'There were two other horses, "The Baroness" and "War Eagle," that were perfectly quiet with me, but would carry no one else without putting them to great trouble, not to say danger. People wondered how I made my horses so showy and cheerful, and yet so gentle and docile. The secret is simple enough. Kindness, sympathy, and patience. I am never unkind to any animal, and they seem to know it.—R. S. CREASY.'

There is one point in the horse's character which the ordinary groom invariably ignores. This is the animal's extraordinary capacity for affection. He *must* love something, man in preference to any other being. In default of man, he will naturally be best pleased with the society of his own kind. He is essentially as gregarious an animal as the sheep, and in the wild state is never found alone. In North America, as is well known, the horses always assemble in large droves, and even in the semi-wild state to which the horse recedes in Bulgaria, the horses obey the same instinct. Mr. H. C. Barkley, in his 'Five Years in Bulgaria,' takes notice of this fact :—

'The great droves of horses and cattle, and flocks of sheep, give the plain a very pretty appearance. As for the horses, they are but little trouble, for in winter and summer they feed themselves on the open plains, and there increase and multiply. They go about in droves of about thirty, with one stallion who acts as master over all, and keeps them in order. Woe betide a young lady who casts sheep's eyes towards a neighbouring drove, or a colt who wishes to enlarge his mind by an interchange of ideas with the young bloods of another family. The vicious-looking old husband and father trots quietly up to the delinquent, and either takes at one bite about a pound of flesh out of its neck, or gives it a kick on the hocks that reduces it to three legs for a week.'

It is sheer cruelty to box up horses after they have done their work, and to keep them from communicating with their own kind.

Were there any other inducement to make man sympathetic with the horse, and to 'gentle' it, as the Americans say, instead of 'breaking' it as we say in England, it is the greater amount of work which can be got out of a horse by treating him kindly. Everyone who is conversant with nautical matters knows how much more work is done by sailors who 'pull with a will,' than by those who merely pull under compulsion, and in this respect horses are very much like men.

Not that horses should be over-indulged or 'spoiled.' They are none the happier for it, but, like spoiled dogs and children, are always wilful and discontented. A ship's crew needs that the captain shall be firm as well as kind, and that his orders must be instantaneously obeyed. It is good that mercy should temper justice, but unless justice be paramount, mercy becomes a cruel weakness. So the horse must never be allowed to feel for a moment that he can have his own way in defiance of his owner's will. With the horse as with man, 'service is perfect freedom,' and both are the happier when this principle is carried out.

In fact, the horse wants to obey man, and if its master will let the animal know his wishes, it will obey him, even though he may not be present. From among many examples of this fact, I select one or two, premising that as some of the personages are still living, the narrator among the number, I employ initials. All the names, however, are given fully in the original manuscript, and the writer is a gentleman of widespread literary fame:—

'When I was a boy, I lived at R—— in Cornwall.

A circus came there, and, a breakage having occurred in some of the iron work, it was repaired by A. M——. The proprietor of the circus went off without paying, and was followed by his creditor, who at last consented to accept a horse in lieu of money. So the horse, a piebald one, such as are often found in a circus, was taken back to R——.

'A few days afterwards, the new owner having to settle an account at some little distance, determined to ride there. All went well until he arrived at his destination, when, seeing a circular flower-bed, with a gravel path round it, the horse took it into his head that the path was a circus, and accordingly dashed into it, and began to gallop round and round in circus fashion.

'Poor A——, a short, stout man, and a very bad rider, clung to the horse's neck, and then suddenly realised the fact that he was enacting the part of John Gilpin. Meanwhile, off flew his hat, strictly according to the part; the enraged gardener held up his spade and tried to stop the horse, and save the flowers, which were being kicked to pieces, and hurled right and left. The worst of it was that the proprietor of the house and garden, Mr. S——, was looking on from an upper window.

'Presently the horse stopped, and A—— hoped to get down. But there was no such luck for him.

The horse had only completed the number of rounds which would bring him to the Calenderer's door, and presently started off again. Not until he had finished the full number of circuits did he stop, and then stood as quiet as a lamb.

'His miserable master, flushed, dishevelled, and full of shame, was then shown into the very room where Mr. S—— had witnessed the adventure. In his confusion he tried to stammer out some sort of apology for the mischief which he had wrought. Mr. S——, however, so far from being angry, begged him to come and repeat the performance as often as he liked, for to see a real John Gilpin was worth all the flowers in the garden.

'For some days afterwards, the horse was safely kept in the stable, but as a fair was to be held at a neighbouring village, C——, the horse was sent, not ridden there, and placed in a shed, while A—— went into the fair to find a purchaser. When he returned, the horse was gone, and so was a window at the end of the shed. Pieces of wood and glass were seen scattered outside, and it was evident that the horse must have leaped through it into a large field.

'On looking through the window, the horse was seen at the farther end of the field, careering round and round as before, to the admiration of a crowd of spectators. No one could stop him until he had finished his task, although four men were offered sixpence each to catch him. The worst of it was that he was so cut and wounded by the broken glass, that he was re-sold for a trifle to the proprietor of the circus.'

' Perhaps the horse was happier in the restricted life to which he had been accustomed, than in the comparative liberty of the road. The account was sent to a nephew of the unwilling John Gilpin, and corroborated by him before it was transmitted to me.

A similar act of conscientious duty on the part of the horse was lately told me by a gentleman whose father witnessed the occurrence.

About sixty years ago the Blandford four-horse mail coach came in as usual, but without coachman, guard, or passengers. It was noticed that the horses galloped up the street at full speed, this being the etiquette at the conclusion of every stage, turned through the narrow archway leading to the inn where they were accustomed to stop, and pulled up without the slightest mishap.

The difficulty of this task can only be understood by those who have undertaken it. To take a carriage and pair through a narrow gateway is by no means easy, and requires considerable practice. In the case of four horses, the difficulty is more than doubled. Yet the horses achieved the task without any help, their memory having served them in lieu of the coachman's guidance.

On inquiry, it turned out that when the horses were changed at the preceding change, and the coach deserted for a time, the team had taken the coach into their own management. A feeble old ostler was at their heads, but he said that when he caught at the reins the horses whisked him out of the way 'like a ninepin,' and never could have stopped until they pulled up at the inn at Blandford.

As in the John Gilpin case, the obedient character of the horse is sometimes apt to give a very ludicrous and scarcely dignified aspect to the adventure in which the animal is engaged.

Some years ago a chaise horse was bought by a Nonconforming minister, and employed in the somewhat staid, though easy task, of conveying him to and fro on his labours. After a few years of such work, the man died, and the chaise was then driven by his widow, who had the misfortune of being stone deaf.

One day a review took place at some little distance from her house at Weymouth, and the old lady insisted on going to see it, and driving the chaise herself.

She was seated in the chaise, looking at the manœuvres, when her horse suddenly bolted, and to her great terror made straight for a cavalry regiment which was on the ground. The horse had heard a sound which he knew, and which she, owing to her deafness, did not hear, and would not have understood if she had heard it.

It was a call from the cavalry bugle, one of the preparatory orders before a charge. The commanding officer saw the state of things, opened out his men right and left, and in dashed the horse to his right place. He took the correct alignment, and stood ready for the next order, which followed immediately.

The cavalry charged, and so did the old lady, the horse being in the proper place, she and the chaise being in the rear. The horse would not leave the regiment, and his mistress would not leave the chaise, and throughout the manœuvres, horse, old lady, and chaise did their part, the horse knowing the bugle calls as well as any of the men, though so many years had elapsed since he had heard them.

When the review was over, the delighted officers thronged round the old horse, and begged to be allowed to buy him and keep him as a regimental pet for the rest of his days. His mistress, however, refused to part with him, and so she kept the old grey, but never again took him within sound of a cavalry bugle.

The history of this adventure was sent to me by a relative of the resolute old heroine, who, in spite of her natural alarm at the unexpected movements, refused to abandon her seat in the carriage, or to part with the old friend who had been her companion for so many years.

This is not a book of anecdotes, and so I will only briefly mention two instances illustrative of the principle of obedience which is so characteristic of the animal, both belonging to the cavalry horse.

At the end of the Peninsular War, a Yorkshire cavalry regiment was disbanded, and the horses put up for sale. The commanding officer, a wealthy Yorkshire gentleman, could not bear the idea that his old companions in battle should pass into the possession of men who would not appreciate them nor understand their ways. He therefore bought them all, and purchased a large paddock, in which they might pass the rest of their lives in honourable retirement.

They had remained in the paddock for some years, when a violent thunderstorm took place. The animals mistook the rolling thunder for the roar of cannon, and the lightning for the flashes of the guns. Their old habits of military obedience returned, and they voluntarily assembled and ranged themselves in battle order.

A somewhat similar instance occurred only a few years ago, and is recorded by the well-known writer who employs the *nom de plume* of 'Rapier' in the *Illustrated Sporting and Dramatic News*. The account appeared in the issue of August 2, 1884 :—

'A soldier friend, who has been quartered in the Transvaal and Basutoland, was the other day telling me of some of his Cape experiences, amongst which one particularly interested me. I am sorry to say that I cannot remember whether it was just after the Zulu or Boer campaign, but at all events it had to do with the 14th Hussars, who were in camp near Newcastle.

'This regiment had brought its horses from India, many of them being stallions, and consequently of high courage. One fine night, disturbed either by flies or the bright moon, a dozen or so broke from the picketing lines, and, careering through the others, caused a regular stampede, so that in a few seconds the whole regiment of horses was loose and galloping wildly towards the river, which was crossed in safety, some hitting on the ford, but the greater part swimming.

'On the far side was a "veldt" or open space,

where the regiment drilled daily, and here a very remarkable thing happened.

'Where there had been chaos and blind uncertainty, perfect order supervened. Forming up into troops and squadrons in their accustomed places, the riderless horses proceeded quietly and steadily to put themselves through a number of evolutions; and it was not until the arrival of the troopers with ropes and bridles that the ghostly pageant—for such it must have looked—broke up and dispersed to the four corners of the earth.'—' Rapier.'

The illustrated newspapers which chronicled the deeds of the Germans and French during their terrible war, recorded several instances in which the riderless, wounded, and even dying horses obeyed the sound of the bugle, and ranged themselves in order after the battle was over.

In answer to some queries which I addressed to him, Mr. Luck kindly sent me the following letter, which arrived too late for insertion in its proper place:—

'Darlington: October 3, 1884.

'My cob was six years old when I took off the shoes in May 1882, and had the thrush in *all* his feet. I am sorry to say that the thrush is not completely cured in his off fore-foot, though we can keep it down, but it returns again and again. I ride and drive him not very long distances, twenty-five miles being the greatest distance he is called upon to do in a day, though he has sometimes done much more. Last week I drove him over twenty miles a day for four days in succession. Our roads are metalled with a stone as hard as granite, and of course in the

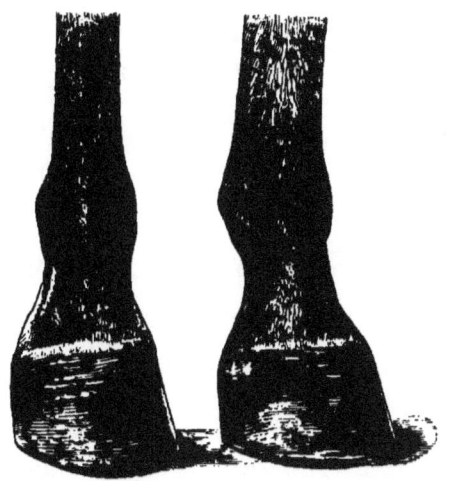

MR. R. A. LUCK'S HORSE, FORE LEGS NEARLY PERFECT.

country there are quantities of loose stones. I enclose you two photos taken fifteen months after the shoes had been removed. Of course the hoofs are much better shaped now.

'Yours, &c.,

'R. A. LUCK.'

The preceding pages were sent to the printer just

before I started for a lecture-tour in America. On the return voyage, one of the fortunate passengers who are never sea-sick, occupied a seat opposite mine at table. We naturally became rather intimate, and I very soon found out that he was specially interested in horses. Being himself a sailor, he was not much of a horseman, but was very fond of horses, and probably knew more of them than he would have done if he had been brought up in a stable.

Among other subjects of conversation was the careless manner in which stables were built, and the insufficient supply of fresh air. Now, as Faraday has shown, the enormous lungs of a horse require more air than would be sufficient for ten human beings, and yet we find that almost all stables are so close that a man feels half-stifled when he enters them.

I happened to mention the fact that the horses which are owned by the North American Indians do not know what a stable means, and even in the coldest weather pass their whole lives in the open air, just as they would do if they were wild (see Colonel Dodge's account of the Indian's horse, quoted on pages 148 and 149).

My naval friend then told me a curious adventure which had just occurred to a horse which he knew well.

Last January (1885) was very cold in Massachusetts, and all traffic was conducted by means of sleighs. On January 30, a horse was being driven through Cambridge, when it took fright and ran away. It upset the sleigh, flinging the occupants into the snow, and then dashed off, banging the over-turned sleigh against all objects that came in its path.

Before the former occupants could pick themselves out of the snow, the sleigh and horse were out of sight. Inquiries were made for them, but without any success, both having vanished as if by a conjuring trick. This happened on a Tuesday morning, and not until the afternoon of the following Friday were the missing horse and fragments of the sleigh discovered.

The animal, urged by the blind terror which sometimes takes possession of a horse, had dashed between two barns. The walls of the barns were not quite parallel, but slightly sloped inwards.

So the horse found himself in a wedge-shaped passage, too narrow, in fact, to allow him to pass through it. When he tried to back out of the passage, he was stopped by the broken sleigh, which was firmly jammed crosswise between the barns.

The weather was very cold, twenty degrees below freezing point, and there was a fierce north-west wind blowing. Unfortunately the barns were built in a north-west direction, so that the passage between them acted as a funnel, and allowed the wind to blow with double fury over the imprisoned animal.

There the poor horse remained, having nothing to eat or drink, no covering, and the furious wind blowing over its unprotected body. It was discovered accidentally, and, strangely enough, seemed little the worse for its long imprisonment. Fortunately, it was a horse that had only lately been imported from the West, and so had not been accustomed to the hot stables of New England. Moreover, it had not been clipped, and, therefore, was wearing the natural greatcoat which is provided for every horse at the end of autumn.

This little history is so valuable in corroborating Colonel Dodge's account of the horse's power of endurance, that I am glad of the opportunity of including it in this book.

On my return home, I found awaiting me a vast mass of correspondence. Among the letters was a long one from Australia, giving accounts of several pet animals. One of these animals was a horse named 'Sam,' who displayed such a talent for 'jockeying' his owner, after the fashion of the American horse mentioned on page 259, that I insert it here, rather than omit it altogether.

The wife of Mr. N., the owner of 'Sam,' is now in

England, and has corroborated the absolute truth of the little histories :—

'"Sam" was a handsome, flighty, high-spirited roan, with black points. He had carried his master faithfully during some ten years of arduous travelling, through rivers and forests, over mountains and trackless country, such as people in England would never dream of, much less think of riding over.

'So far, all had gone well, but on one unlucky day "Sam" lost a shoe, and before any blacksmith's forge could be reached, he had gone so lame, that it was necessary to turn him out to grass. His owner, Mr. N., procured a substitute, and "Sam" was left for several months undisturbed in his paddock.

'At last, Mr. N. caught him, and rode him some eight or ten miles to the blacksmith to have him shod, as a preliminary to resuming the usual day's work. They reached the forge comfortably enough, and the horse was re-shod. But they had scarcely proceeded a couple of hundred yards, when "Sam" fell dead lame, his head bobbing up and down, and his whole demeanour indicating the acutest agony.

'Of course, common humanity required that the day's expedition should be given up, and there was no choice for Mr. N. but to retrace his steps. But, O Miracle ! No sooner was "Sam's" head turned homewards, than the lameness vanished.

'Three times was the experiment tried with precisely the same results, when it occurred to Mr. N. to compromise the matter, and by taking " Sam " an indirect way, he made his calls, and finished his journey to the satisfaction of all parties. " Sam " was evidently aware that he was indebted to his lameness for his holiday, and had feigned lameness for the sake of another respite from work, but was not clever enough to carry out the deception to its full extent.

'Without the least approach to vice, " Sam " was masterful and overbearing with regard to other horses. On one occasion, four beside himself were running in the same paddock, and as it was in the winter time, we were accustomed to throw down five large bundles of hay—one for each horse.

'Now, " Sam " made it a point of honour to drive the other horses from their food. By constant vigilance, he could manage to keep guard over three heaps of hay beside his own, but the fourth was beyond his powers, and evidently cost him great trouble of mind.

'One day, however, a sudden inspiration seized him. As soon as the hay was brought, he dashed at the obnoxious heap, rolled it in the mud so as to spoil it, and then returned complacently to his usual position.

'I have also seen "Sam" when in want of water go

to the pump and work the handle with his mouth, evidently knowing by personal observation that this was the mode by which water was obtained.

'His companion was an exceedingly intelligent and beautiful white pony, and the two were so clever at opening my ordinary fastening to the garden-gate, that a complicated arrangement of straps had to be devised in order to keep them out.

'Under the pump was a large tank, out of which they were accustomed to drink. On one occasion a washing-basin had been left in the tank. The two horses undid the fastenings of the gate, let themselves into the garden, and proceeded as usual to the tank. They objected to the presence of the unaccustomed basin, took it out of the tank and put it on the grass before they would drink.'

# INDEX.

### ABY
ABYSSINIAN hunting, 145
Accidents, 118
Aloes, 98
Anchitherium, 14
Ankle of horse, 19
'Apollo,' 163
Astley, Mr. A. F.'s horses, 158, &c.
Attitude, natural, 301
Australian mare 'Fly,' 315

BAKER, MR. WHITMORE, 164, 168
Barclay and Perkins, 228, 298
'Baroness,' 316
Bearing-rein, 199, 205
— American, 209
— diseases caused by, 220
— overhead, 209
— and stumbling, 229
— summary, 240
Bedouin-rein, 207
Bellows, Mr. J., 83
Bergh, Henry, 220
Blacking hoofs, 93
Blandford coach, 321
Blinkers, 244
Boot, fashionable, 125
Bowditch, Mr., 122
Bowley, Mr. Samuel, 275

### COM
Brakes and bearing-reins, 226
Brierley, Dr., 180
Bristol carters, 231
Brushing, 196
'Buffers,' 214
Bulgarian horses, 317
Burnaby, Col. Fred., 156
'Burr' bit, 221

CALKS, 116
— in America, 121
— effect on muscles, 126
Carpus, 11
Cavalry horses, 311, 322
Caves, men of the, 17
Character of the horse, 248
Charlier shoe, 134
Chapin, Mr. S., 73
Check-rein, the, 206
Children, restless, 304
Clever comrades, 333
Clicking, 196
Climbing, 24
Clipping, 277, 282
'Clips,' 80
Coach, Blandford, 321
Coffin-bone, 15
Coloured diagrams, 21
Comrades, clever, 333

## INDEX.

### COR

Coronary ring, 30, 44, 92
Cracknell, Mr. E., 232
Cressy, Mr. R. S., letter of, 315
Crib biting, 304
'Cropping,' 246
Cruelty the cause of vice, 308
'Cruiser,' 309
Crupper, the, 276
Curb, the, 202
Curiosity of the horse, 249
Cutting, 196

DEAFNESS, 247
Derby, the, 4
'Dewdrops,' 60
Diagrams, coloured, 299
Diseases, caused by bearing-rein, 220
— caused by shoe, 197
Dixie, Lady Florence, 25
'Docking' the tail, 267
'Dolly' and loose stones, 60
Door of stable, 302
Drivers of engines, 224
— of horses, 225
Drugs, 96
Duty, sense of, 319
Duties, reciprocal, xvii.

EAR, the horse's, 247
Edinburgh horses, 230
Egyptians, ancient, 206, 279
Eight-hoofed horse, 18
Elbow of horse, 12
Endurance of horse, 330
Engine-drivers, examination of, 224
Exmoor ponies, 24
Expansion of hoof, 51
Eye of the horse, 260
— professional, 196

### HON

FAILURES, apparent, 192
Falkland Island horse, 30
Farriers, directions to, 57
'Fashion,' 200
Fear, rule of, 307
Femur, 18
Fetlock, angle of, 33
— section of, 40
Fibula, 19
Floors, sloping, 299
Flower, Mr. E. F., 199
'Fly,' 315
Foam-flinging, 217
Foot, human, 142
'Foundered' horses, 64
Frog, the, 33
— paring the, 66
— pressure of, 67
— value of, 68

GAG bearing-rein, 207
'Gentling,' 317
Gerry, Mr., letter of, 133
Gill, Mr. S., 220
Gilpin, John, 319
Grazing, mode of, 293
Groom, the kindly, 311

'HAMMEL,' 60
Happy thoughts, 138
— more, 257
Hartmann's safety-pad, 74
'Haw,' the, 263
Head, Sir Francis, 234
— tossing of, 217
Heels, opening, 61
Helps, Sir Arthur, 218
Hilly roads, 230
Hip-bone, injury to, 303
Hipparion, 14
Hock, 19
'Hogging' the mane, 265
Honour, a point of, 332

# INDEX. 337

## HOO

Hoof, the, 22
— alteration of shape, 188
— and foot, 1
— difference in, 179
— internal, 38
— loosened, 115
— offices of, 27
— overgrown, 30, 193
— plasticity of, 171
— of wild horse, 34
Horse and ass, 265
Horse-clothing, American, 278
Horse, endurance of, 329
— ferocious, 312
— home of, 191
— without man, xvii.
Hot-fitting, 86
Houghing, 19
'Humane,' 174
Humerus, 10


'IMPETUOSITIES,' 230
Impostor, an, 258
Imprisoned horse, 329
Indian horses, 148
India-rubber tires, 70
Infected provender, 200
Interfering, 196
Irreclaimable savages, 309
Irreligion, 167
Italian horses, 181


JACKDAW, an unhappy, 286
'Jockeying' the driver, 258
John Gilpin, 319
J.P., a puzzled, 267


KETTLE-DRUMS and Rarey, 250
Kindness, rule of, 307
'Knee' of horse, 11

## ORL

LAMENESS, sham, 331
Laminæ, horny, 29, 45, 48
— vascular, 45, 124
Ligaments of neck, 211
Linea alba, 48
Llewellyn, Dr., 158
Lockjaw, 270
Luck, Mr., horse of, 326
Ludgate Hill, 79, 118
Lungs of horse, 287
'Lurry' horses, 77


'MACCABÆUS,' 4
McIlquham, Mrs., 270
Man without horse, xvii.
Manchester roads, 78
Mane of the horse, 264
Manger, electric, 305
— height of, 302
Manifesto, important, 235
Masters and men, 228
Mayhew on clipping, 282
Mechanism of walking, 53
Memory of horse, 323
Men of the caves, 17
Metacarpals, 15
Metatarsals, 20
'Mill, going through the,' 208
Monday morning, 293
More happy thoughts, 257
Mustangers, 23


'NATURE,' 201
Navicular bone, 41
Neck, ligaments of, 211
'Nicking' the tail, 272
Nictitating membrane, 263


OINTMENTS, 95
Olecranon, 11
'Orlando,' 4

Z

## OVE

Over-feeding, 293
Overgrown hoofs, 30, 193
Overhead-rein, the, 209

Pad, safety, 74
Palmerston, Lord, 234
Parkyns, Mansfield, 145
' Pax-wax,' 211
Pastern, 15
Pelvis, 18
Phalanges, 12
Pharaoh's chariot, 206
Ponies, Exmoor, 24, 153
— Indian, 148
' Pricking' a horse, 105
' Prince,' 182
Professional Eye, the, 198
Pulling ' with a will,' 318

Racing, Indian, 151
Radius, 10
Rails, sloping, 300
Ransom, Mr. G., invention of, 242
Rarey's system, 249, 309
Reciprocal duties, xvii.
Reclaimed savage, 313
Review, a, 322, 325
' Roaring,' 237
Runaway horses, 233
' Running-rein,' 4

' Sam' and his tricks, 331
Savages, irreclaimable. 309
Savage, a reclaimed, 313
' Scaling' birds, 261
Scapula, 9
Scotch girls, 144
Sea-shoes, 81
Sesamoid bones, 42
Shaving, 270, 284
Shaw, Mr., horse of, 234
Shoe, cause of diseases, 197

## TAM

Shoe, Charlier, 134
— Clark, 128
— Goodenough, 127
— jointed, 128
— object of, 101
— parflêche, 103
— screw, 130
— straw, 103
— unilateral, 106, 132
— weight of, 113
Shoeing, hot, 86
Shoe-nails, 104, 108
Shoes, sea, 81
Shoulder-blade, 9
' Showing off' engines, 222
— horses, 223
Sidney, ' Book of the Horse,' 209
Singeing, 277
Skeleton of horse, 3
Slitting birds' tongues, 261
'Smith, Mr. Herbert, 185
Soldiers, 311
Soldier, the old, 322
Sole, the, 36
— paring the, 58
Spikes, 76
Splint bones, 17
Spoiling horses, 318
Spring of railway carriage, 35, 71
Stables, typical, 289
Stable windows, 291
' Stafford,' 309
Staffordshire roads, 195
' Stella,' 164
Stifle-joint, 19
Stock, leathern, 279
Stomach of horse, 292
— of ox, 292
— enlarged, 295
' Stopping,' 95
Stumbling, 229

Tamplin, Mr. C. H., horses of, 253

Tarsus, 19
Tendons, 42
'Throwing back,' 18
'Thrush,' 62
Thunder and lightning, 324
Tibia, 19
'Tommy,' 161
Tongues partially severed, 237
Tossing of head, 217
Traction engines, 223
Training, 297
Tram horses, 174
Trough-water, height of, 302

ULNA, 10

VARNISH, natural, 92
Ventilation, 287
Veterinary surgeons' manifesto, 235
Vice caused by cruelty, 308
'Vices, stable,' 303

WALKING, mechanism of, 53
Wall, the, 29
'War Eagle,' 316
Water, purity of, 299
Watering horses, 295
— in America, 296
— in Turkey, 296
Waterton, Charles, 140, 255
'Weaving,' 303
Welcome, a horse's, 310
Williams, Mr. Theodore E., 182
Windows in stables, 291
Wing clipping, 285
Wolf, mode of ' weaving,' 303

XENOPHON on horses, 189

YORKSHIRE regiment, 324

ZEAL and discretion, 228

www.ingramcontent.com/pod-product-compliance
Lightning Source LLC
Chambersburg PA
CBHW020246240426
43672CB00006B/652